Essays

On The Global Economy
&
World Financial Markets

Essays

On The Global Economy & World Financial Markets

Ranga Chand

Copyright © 2021 Ranga Chand

All Rights Reserved

ISBN: 9798502180689

Imprint: Independently published

To Sylvia and Jason

CONTENTS

Preface xi

PART 1 - THE YEAR 2009

1 What's in Store for 2009? Some Observations 1
2 Canada's Economy Hits A Brick Wall 4
3 Central Banks Embrace Quantitative Easing – But Will It Work? 8
4 Can China Lead the World Out of Recession? 12
5 US Recovery Still Months Away 17
6 The Battle against Deflation Still to be Won 21
7 This Rally Looks Unsustainable: But it's Not All Bad News 27
8 After The Recession, An Erratic And Jobless Recovery 33
9 What's Driving the Surge in Global Equity Markets 38
10 Why a Weaker Canadian Dollar Matters 44
11 World Economy Slipping into Deflation Territory 48
12 What's in Store for 2010? Some Observations 53

PART 2 - THE YEAR 2010

13 Global Stock Market Performance in the 2000s 58
14 US Economy Faces Precarious Recovery Path 63
15 Greece's Agony – An Economy on the Brink of Disaster 68
16 A Stronger Yuan No Cure for an Ailing Us Economy 72
17 Soaring G7 Government Debt and the Interest Rate Paradox 77

18	From Recession to Recovery in the G7 – An Early Assessment	83
19	The Bears Are Circling Again	88
20	The Curious Case of Falling Bond Yields in G7 Economies	93
21	From Recession to Depression – The Wrenching Case of Ireland	99
22	The Federal Reserve and QE2 – A Hazardous Strategy	105
23	Germany Emerging as Europe's Powerhouse	111

PART 3 - THE YEAR 2011

24	Canada's Economy Faces Challenges	117
25	Inflation and the G7 Economies	123
26	The Feds Zero-Rate Interest Rate Policy – It's Time for a Change	128
27	Recession and Recovery in the Eurozone – A Tale of Diverging Paths	132
28	The Great Recessions Impact on Government Finances in the G7	137
29	Credit Ratings and Government Debt – A Tale of Woe	142
30	The Faltering Recovery in G7 Economies	147
31	Greece on Fast Track to Default	152
32	The Looming Break-Up of the Eurozone	157

PART 4 - THE YEAR 2012

33	E7 Growth Performance Trumps G7	164
34	Canada – The Mouse That Roared	169
35	Central Banks & Quantitative Easing – Time to Change Course	173
36	Mounting Debt Levels to Overwhelm G7 Economies	177
37	Current Policy Paths Dooms the Eurozone	180

PART 5

ARTICLES PUBLISHED BY THE GLOBE AND MAIL

(Canada's National Newspaper)

I	G7 Revenues and Spending: Mind the Gap	186
II	The Price of Debt: Higher Borrowing Costs	188
III	Waiting for Sustainable Growth	190
IV	Rise of the Emerging Seven Reshaping the Global Economy	192
V	It's Time for a New Approach to Stimulus	194
VI	IMF Projections Question G7's Commitment to Tackling Debt	196

About the Author	199

Preface

The 2007-2008 financial crises and the Great Recession that followed was a global event that caused great upheaval in the world economy. The crisis, which started in the US with the sub-prime mortgage debacle and the bursting of the housing bubble, triggered a credit crunch that reverberated around the globe and came close to toppling the world's financial infrastructure.

The developed countries were particularly hard hit as GDP dropped sharply in all the major Western economies. With demand plunging and job losses mounting, the unemployment rate soared. Compared to their pre-crisis levels, the rate doubled in several countries including Ireland, Spain and the United States and tripled in the Baltic countries of Latvia and Lithuania.

Tens of thousands of businesses, both large and small from all economic sectors, went bankrupt. For households, as property prices tanked millions of homeowners found themselves in negative equity, owing more on their mortgages than their homes were worth.

When Lehman Brothers, the behemoth Wall Street bank, collapsed in September 2008 it was a watershed moment. Credit markets froze and liquidity virtually vanished as banks were reluctant to lend for fear of not being repaid which gummed up the economic machinery. As panic rippled through the system and investors headed for the exits, global stock markets took a battering. By the end of 2008 trillions of dollars in value had been wiped out.

This book brings together a series of essays that the author wrote for his corporate clients over the period from 2009 to 2012. The articles analyzed the impact and fallout that the Great Recession had on the global economy and critically examined the efficacy of the policy responses of governments and central banks as they tried to steady the global economic ship.

Also included are a number of posts by the author that were published in the Globe and Mail, Canada's premier national newspaper.

PART 1

The Year 2009

1

What's In Store For 2009?
— *Some Observations* —

(January 2009)

Credit Crunch to Abate, but Slowly

The global economic contraction that is now underway is set to intensify with world output projected to fall sharply in the first two quarters of 2009. Despite steep reductions in interest rates by the world's central banks and several multi-billion dollar financial bailout packages aimed at shoring up the banking sector, a recovery is still not in sight and credit markets remain dysfunctional.

The financial crisis, which started in the US with the sub-prime mortgage debacle and the bursting of the housing bubble, triggered a credit crunch that continues to reverberate around the globe. As banks struggle to repair their balance sheets and write down their toxic debt holdings, credit lines are being reduced and lending standards tightened. The bottom line is that banks are contracting the amount of lending.

Although interbank lending rates have come down significantly from their record highs set last fall following the bankruptcy of Lehman Brothers, they still remain elevated relative to the Federal

Reserve base rate. The spread between the 3-month Libor rate and the Fed funds rate is still north of one percentage point whereas historically the difference between the rates has averaged about one-tenth of one percentage point. With credit conditions continuing to remain tight, a pick-up in economic activity is still a distance away.

Fighting Deflation is Today's Priority; Inflation is Tomorrow's

As global demand continues to spiral down and inflation rates tumble around the globe, we are headed for a period of deflation in the short term. The Federal Reserve and other global central banks will remain in crisis mode as they battle to prevent deflation from taking hold. Having virtually used up their ammunition on the interest rate front, a strategy that failed to stoke the engines of economic growth, the focus will now be on quantitative easing; a euphemism for basically printing money.

With the monetary spigots now fully turned on and set to soak the world economy with liquidity, a lengthy bout of inflation looms in the future. But that's tomorrow's problem; ensuring that deflation doesn't take hold remains today's challenge.

Bear Market Rallies to Test Investors Resolve

Expect the major stock markets to grind their way higher, but be prepared for large volatile swings along the way. Don't be surprised if stock markets rebound strongly during the first few weeks of the new Obama administration. The 'feel good' factor and pure 'animal spirits' will boost market confidence but the Obama bounce will occur against a background of deteriorating economic conditions, falling corporate profits, and surging unemployment.

While equity markets this year will likely be punctuated with short 'bear market' rallies, it is worth keeping in mind that for investors with a longer-time horizon – 10 years or more – now may be a good time to start gradually easing into the market given depressed stock prices.

Declining Living Standards to Challenge Western Economies

The major focus for western economies in 2009 will be a combination of economic contraction, soaring unemployment, and the fall in standards of living. In terms of growth, the American and British economies will take the biggest hit with GDP expected to decline by at least 2%. Declines in the Eurozone area, Japan and Canada are projected to range between -1% and -2%.

With both monetary and fiscal policy continuing to be extremely expansionary, we should start to see a recovery in output towards the end of 2009. However, whether the recovery will be sustainable will depend on the speed with which credit markets return to 'normal' and the extent to which the ease of access to loans by businesses and consumers resumes again. At the time of writing, the prognosis does not look very encouraging – there continues to be a lack of confidence and trust in the financial sector.

Protectionism Set to Re-Appear

With trade volumes expected by the World Bank to shrink this year by 2% – the first outright decline since 1982 – expect calls for protectionism to start rising. The heat will be on for governments worldwide to shore up their domestic economies and protect jobs at home.

Faced with great economic uncertainty both domestically and internationally, it will be a huge challenge for the political leadership to keep their markets open particularly at a time when the voting public is beginning to question the gains from globalization.

2

Canada's Economy Hits A Brick Wall

(February 2009)

Economic Indicators Turn Down Sharply

After managing to avoid the global downturn for much of 2008, the Canadian economy has now crashed headlong into a brick wall. Incoming data for all the key indicators including retail sales, home sales, residential construction, manufacturing shipments, new orders, and exports uniformly shows that the wheels have come off the Canadian economy.

Any doubts about the validity of the notion that the downturn was intensifying were dispelled when Statistics Canada reported last week that the economy had shed 129,000 jobs in January, the largest monthly fall in employment since 1976. The unemployment rate jumped to 7.2% from 6.6% in December and is up sharply from its 33-year low rate of 5.8% set back in January 2008. Interestingly, although January's 598,000 job loss in the United States was a real eye-opener, when one adjusts for the difference in populations between the two countries, Canada's job loss was twice as bad, implying a comparable loss in employment of 1.2 million in the US.

With the Canadian economy now caught in the throes of a recession, attention has turned to finding answers to the following questions: how long will the downturn last, how deep will it be, and, once the economy hits bottom, how robust a recovery can one expect.

The general consensus is that the recession is expected to last between 3 and 4 quarters with GDP falling by a cumulative 2 1/2 to 3 percentage points from peak to trough. Most analysts also expect that as the combination of expansionary monetary policy and the recently announced fiscal policy measures contained in the Federal Government's January Budget works its way through the system, an economic recovery will be underway by the fourth quarter of 2009. The consensus view is that GDP growth will range between 1.5% and 2% next year.

Bank of Canada – Right on Monetary Policy Response but Wrong on Recovery Forecast

Contrary to the consensus view, the Bank of Canada expects the recession will be short lived lasting 3 quarters and should be over by the end of the second quarter of this year. However, what is startling is that the Bank expects a V-shaped recovery with the economy turning around strongly from its low point. In its latest Monetary Policy Report, the Bank is forecasting annualized growth of 2% in Q3 and 3.5% in Q4. The pace is also projected to pick up strongly in 2010 with growth expected to average 4.5% in the first six months of the year and 4.7% in the second half.

The Bank's GDP growth forecast of 3.8% for all of 2010 is more than double the 1.6% pace predicted by the IMF in its latest world outlook. The Bank's forecast of a short recession followed by a strong recovery in 2010 has failed to inspire much confidence in the markets which continue to trade sideways.

Fortunately, while the central bank's forecast of GDP growth is wildly optimistic and stretches credibility, its stance on monetary policy has been correct. Responding to the ongoing global credit crisis and the deepening global downturn, the Bank has moved expeditiously to shore up the Canadian economy by aggressively lowering interest rates and ensuring that the financial system has ready access to liquidity.

Will Expansionary Monetary and Fiscal Policy Response Rescue the Canadian Economy from Recession?

In a 'normal' recession, when the economy is performing well below full capacity, expansionary monetary and fiscal policies will, albeit after a time lag ranging anywhere from 6 to 18 months, boost aggregate demand. However, the Canadian economy is experiencing anything but a normal economic downturn. Like all other countries, it is now feeling the full blast of the credit crisis that has gripped the global financial system for the past several months.

As a direct consequence of the credit squeeze the global economy is undergoing its first synchronized recession since the 1930s. As worldwide demand collapses, export dependent economies like Canada are being hit hard which in turn is forcing companies to cut production and lay off workers.

Moreover, Canadian consumers are saddled with a huge debt burden – the ratio of debt to disposable income is at a record high of 130% – and as the recession deepens the default rate on credit cards is starting to rise. Unsurprisingly, consumers are shifting their focus away from spending to paying down debt and as a consequence the savings rate is rising. As households repair their balance sheets, this deleveraging process will result in slower growth.

On the monetary policy front, it is clear that the slashing of interest rates by the central bank has so far failed to ignite demand. The economy continues to contract, consumer spending is falling sharply, job losses are mounting, and the unemployment rate is quickly rising.

The fiscal stimulus measures contained in the Government's budget will help cushion the downturn but whether it will lead to a sustainable recovery where each quarter shows positive growth is debatable. It clearly is going to take time to get the spending plans up and running and, given the weak state of the economy, the impact of tax cuts is unlikely to boost spending in any meaningful way.

The Bottom Line

Although expansionary fiscal and monetary policy measures may help mitigate a further deterioration in economic conditions, 2009 unfortunately promises to be a year of economic hardship for Canadians and the Canadian economy. Moreover, as we take a look beyond the current economic crisis, it seems that the recovery when it comes will likely be slow and arduous. Given their debt loads, consumers will either accept or be forced into accepting a new frugal lifestyle. While painful in the near term, this new age of consumer frugality will pave the way for a more balanced and healthy economy down the line.

But, unless the global financial system stabilizes and credit starts to flow unimpeded again, economic policy measures aimed at boosting demand are unlikely to lead to sustained growth.

3

Central Banks Embrace Quantitative Easing

— But Will It Work? —

(March 2009)

Credit Crisis Renders Monetary Policy Impotent

Central banks around the world have been slashing interest rates at a furious pace over the last year or so in an attempt to unclog credit markets that had seized up following the collapse of the US housing market and the implosion of the sub-prime mortgage market.

Base rates now range from 0.1% for the Bank of Japan to 0.25% for the Federal Reserve to 0.5% for both the Bank of England and the Bank of Canada. Yet the moves to jump start their moribund economies by pursuing what amounts to essentially a zero interest rate policy have clearly failed to stem the financial crisis.

Credit markets remain dysfunctional and trust – the glue that holds the financial system together – has all but tanked. Banks continue to be leery of making loans, even short-term ones, for fear that they may not be paid back. Moreover, repairing and purging their balance sheets from toxic assets remains the primary focus of several financial institutions.

Predictably, as credit conditions have tightened and the availability of credit has been reduced, the crisis has spilled over into the real economy. The global economy is now caught in a nasty downward spiral and real GDP is contracting at an alarming rate in several major economies. Moreover, job losses are beginning to surge bringing with it the spectre of protectionism and a renewed but still incipient threat of nationalism.

Money Creation via Quantitative Easing

Having run out of conventional weapons to fight the deepening global recession, monetary authorities are now turning to the electronic printing presses in a last ditch attempt to shore up demand and prevent a deflationary spiral from taking hold. With the failure of record low interest rates to ease credit conditions and re-ignite demand, several central banks are now planning to revive growth via quantitative easing.

The US Federal Reserve, the Bank of Japan, the Bank of England, and the Bank of Canada have taken, or are planning to take, radical steps to increase the money supply by printing money in order to buy up billions of dollars of securities including corporate debt and government bonds from financial institutions. Essentially, the central banks will pay for these assets by electronically creating new money and crediting it to the reserve balances of the financial institutions. This way armed with 'excess reserves' commercial banks will have an incentive to work off these excess reserves by increasing their loan book.

The Goals of Quantitative Easing

By pursuing a policy of quantitative easing, the central banks hope to achieve two goals. *Firstly*, by boosting the money supply the hope is that this will lead to an easing of tight credit conditions and a concomitant increase in business and consumer loans. *Secondly*, by buying up corporate and government bonds, quantitative easing will lead to higher bond prices and lower interest rates in the marketplace.

As money starts to flow through the economy again this will lead to higher spending and hopefully stop a potentially destructive fall in prices from taking hold. The key question of course is will all this boosting of the money supply through quantitative easing actually lead to a pick-up in economic activity? Or will it merely set the stage for a sharp rise in inflation in the near future?

Will Quantitative Easing Work?

Given that the world economy is experiencing a severe recession where consumers and businesses are saving instead of spending, it is unlikely that economic activity will pick up in a meaningful way in the near future. Consumers remain highly leveraged and corporate bankruptcies are on the rise. The central banks are assuming that by shoring up the balance sheets of financial institutions (through buying up billions of dollars worth of securities in exchange for money) the latter will start to lend to borrowers thereby putting more liquidity into the economy.

The trouble is that with the recession deepening and a growing number of businesses going under this means that the bad debts on bank balance sheets will rise this year. And as banks increase their provisions for loan-losses, credit conditions will inevitably tighten and further curtail their ability to lend. Moreover, despite the increase in the money supply, a sharp slowdown in the 'velocity of circulation' of money – the rate at which money changes hands – which typically occurs during recessionary periods because of hoarding could easily stymie the effect of any quantitative easing.

On the inflation front, it is doubtful that quantitative easing will lead to a surge in inflation – at least in the near term. The World

Bank, in its latest projections for the global economy, predicts that growth will be at least 5 percentage points below potential. Moreover, the Bank anticipates that global industrial production by mid-2009 could be as much as 15% lower than 2008 levels. With demand imploding virtually across the board, there is a very real danger that deflation might take hold as firms strive to rein in supply to meet the dramatically reduced level of demand.

The Bottom Line

There is no guarantee that a policy of pumping money into the economy through quantitative easing will arrest the deepening recession nor can one confidently assert that it will lead to a resumption of growth. However, a failure to aggressively pursue quantitative easing could well doom the global economy to a multi-year economic slump with all the human misery this would entail. Talk about policy makers being caught between a rock and a hard place.

Not surprisingly, the rate cuts by the central banks instead of boosting aggregate demand are having little or no effect. Spooked by the recession and surging unemployment, consumers have switched gears and are now intent on paying down debt. As a consequence, savings rates are rising and this is further strangling economic activity. Until this deleveraging process runs its course – which could be many months away – and debt levels become more manageable, economic growth will amount to a slow crawl. Not a very encouraging outlook, but there you have it.

4

Can China Lead The World Out Of Recession?

(April 2009)

Global Downturn Deepening

Despite vigorous policy efforts on the monetary and fiscal front by governments around the world, economic activity is falling precipitously. Global GDP fell by an unprecedented 5% on an annualized basis in the fourth quarter of 2008 and incoming data suggests that the fall in growth in the first quarter of this year is likely to be even more dramatic. Moreover, world trade is also contracting sharply as falling demand and the drying up of trade financing resulting from the credit crunch is further reinforcing the rapid spread of recession.

According to the International Monetary Fund's latest March 2009 forecast, global activity is expected to decline by 0.5% to 1% this year. As a group, growth in the G-7 is projected to contract sharply falling between 3% and 3.5% with little chance of a recovery taking hold before mid-2010. Growth in the emerging economies is projected to average 2%, well down from the 6.1% pace registered in 2008.

China's Economy – A Snapshot

China's economy has expanded rapidly over the past 30 years posting an average annual rate of growth of around 9%. This has pushed up its share of world GDP from less than 1% in 1977 to 6% in 2007. The country is now the world's third largest economy and, notwithstanding the global recession, it is slated to surpass Japan and emerge as the world's second largest within the next 5 years.

Despite the country's emergence as an economic superpower, income levels remain well below those of Western economies. According to the World Bank, GDP per person in China was $2,500 in 2007 whereas it ranged from $35,000 and $46,000 in the G7 economies.

China's Growth Rate Down but Not Out

China's export-driven economy has not been immune to the global slump. After expanding by 13% in 2007 growth has slowed sharply, falling to 6.8% in the fourth quarter of last year. With the collapse in global trade severely impacting the country's export sector, which fell by 26% in February 2009 from a year earlier, thousands of factories producing toys to textiles to steel have shuttered their gates and an estimated 20 million migrant workers have lost their jobs.

To counter the global economic slump and stave off social unrest, the authorities are switching their attention from the external market to the domestic one. However, unlike major Western economies which are faced with both a credit crunch and deepening recessions, China has to deal with only one of these challenges; decelerating growth brought on by the collapse in world trade. To boost domestic demand and offset the sharp decline in exports, the Chinese government has launched a $586bn stimulus package that will focus on infrastructure projects and social spending.

As government spending is being ramped up, the pace of bank lending has also picked up and the economy is showing tentative signs of having turned the corner. This has led to optimism in some circles that China may be the engine that will pull the world economy out of recession.

The question of course is how plausible is this. Despite China's stellar growth performance over the past few decades, one look at the country's economic structure suggests that for China to play a pivotal role as the locomotive to the world economy is premature.

The Unbalanced Structure of China's Economy

As can be seen from Table 1, China's economy has become unbalanced over the years. With its emphasis on an economic strategy that has focused on the export and investment sectors, the country's share of exports has risen from 5% GDP in 1977 to 41% in 2007. (The share of imports has also increased going from 4% to 31%). Underpinning the growth in the export sector, capital investment spending has also zoomed. Gross capital formation now accounts for 43% of the country's GDP up from 28% in 1977. On the other hand, during this time period consumer spending, as a share of GDP, has declined sharply falling from 64% in 1997 to 36% in 2007.

TABLE 1
CHINA
GDP by Type of Expenditure
(Percentage Shares of GDP)*

	Consumer Spending (C)	Investment Spending (I)	Government Expenditures (G)	Exports (X)	Imports (M)
1977	64	28	8	5	4
1987	50	36	14	16	16
1997	45	37	14	21	17
2007	36	43	14	41	31

The Formula for Expenditure GDP = C+I+G+(X−M)
*The breakdown in shares of GDP may not add up to 100 percent due to statistical discrepancies

Source: United Nations Statistics Division

In contrast, it is interesting to look at the evolving structure of the US economy over the past 30 years. Compared to China's economy, consumer spending as a share of GDP has increased from 63% in 1977 to 71% in 2007 (see Table 2). Although the share of exports has also increased during this period rising from 8% to 12% of US GDP, the share of imports has nearly doubled to 17%.

	Consumer Spending (C)	Investment Spending (I)	Government Expenditures (G)	Exports (X)	Imports (M)
1977	63	21	17	8	9
1987	66	19	18	8	11
1997	67	19	15	12	13
2007	71	18	16	12	17

TABLE 2
UNITED STATES
GDP by Type of Expenditure
(Percentage Shares of GDP)*

The Formula for Expenditure GDP = C+I+G+(X−M)

*The breakdown in shares of GDP may not add up to 100 percent due to statistical discrepancies

Source: United Nations Statistics Division

In 2007, according to the United Nations, total household spending globally amounted to $32.2 trillion. US consumers were the most prolific spenders accounting for $9.7 trillion (30%) of this total whereas Chinese households accounted for only $1.2 trillion or 4%.

The US exported $1.6 trillion of goods and services in 2007 but imported $2.4 trillion for a deficit of $800 million. By comparison, China ran a surplus of about $300 million exporting $1.4 trillion and importing $1.1 trillion worth of goods and services.

So, in a nutshell, China's economy has been over-producing and under-consuming whereas the US has been under-producing and over-consuming.

The Bottom Line

Any expectation that China can pull the world economy from its deep recession is grossly misplaced. Although China is projected to post the highest growth rate amongst the major economies in the near term, given its still small weight in the world economy, it will only make a marginal, albeit helpful, contribution to overall growth.

The Chinese economy is heavily dependent on exports and is therefore very exposed to any downturn in export markets. Moreover, there is a long way to go before Chinese consumers will have the earnings capacity to replace US households as the main driver of consumer spending in the world economy. The recovery of the world economy will depend on a revival of the United States and other advanced economies.

Unfortunately, until the global banking system stabilizes and lending resumes unimpeded in the United States and other Western economies, the world economy faces a prolonged period of economic dislocation and sluggish growth.

5

US Recovery Still Months Away

(May 2009)

Investor Confidence Misplaced

Investor confidence has been rising in recent weeks amid signs that the downturn in the US economy is easing and that momentum for a recovery is starting to build. Indeed, the stock market has rebounded strongly over the past two months, consumer confidence has been rising, and credit markets have started to thaw. However, the recent string of downbeat data suggests that this optimism is misplaced and that the so-called 'green shoots' of recovery are beginning to look more like weeds.

After posting modest back-to-back gains in January and February, retail sales numbers for March and April turned down again and re-enforced the fact that consumers are in serious retrenchment mode. Faced with massive job losses – 5.7 million since the start of the recession in December 2007 – and wages that are either stagnating or declining, Americans are struggling to reduce their debt levels to more manageable levels. As they batten down the hatches and curb spending, savings rates have shot up. According to the US Department of Commerce, personal saving as a percentage of disposable personal income has increased sharply over the past few months, rising from 0% in April 2008 to 4.2% in March 2009, and is further dampening consumer spending.

The latest housing report also confirmed that the sector remains in a deep slump as US housing starts plunged 12.8% to a record low 458,000 units in April. Moreover, house prices are also continuing to fall. The S&P/Case-Shiller 20 Home Price Composite Index fell 1.9% during February and house prices are now 18.6% lower than they were in February 2008. Since peaking in May 2006, home prices have fallen by a staggering 29.9%. What's more, despite the sharp fall in average house prices, relative to rents and incomes, house prices still remain above historical norms and may well fall further before stabilizing.

Economic Policy on Steroids

In trying to boost aggregate demand and jump start the economy, policy makers have been very aggressive on both the monetary and fiscal policy fronts. The Federal Reserve has slashed interest rates to record lows and with rates now effectively at zero it has run out of options on this front. Not to be outdone, the Fed has opened up another line of attack and has initiated a blast of quantitative easing aimed at boosting the money supply. By buying up government and corporate bonds, the hope is that this will lead to lower long term interest rates and thereby provide a needed boost to the economy.

Washington has also been ramping up government spending and when one factors in the fall in tax revenues caused by the deep recession, the Wall Street bailout, and various fiscal stimulus measures aimed at shoring up the economy, the budget deficit is set to explode. The White House now expects the deficit to quadruple to $1.8 trillion this year which will push up the deficit to an alarming 12.9% of GDP. To date, however, despite throwing everything including the proverbial kitchen sink at the US economy, the economy remains in a funk.

From a policy perspective there is little more that the Obama administration can do to encourage consumers and businesses to borrow and spend to boost growth. While pumping vast sums of money into the system will undoubtedly lead to a pick-up in growth for perhaps a few intermittent quarters, there is a real danger that the recovery, when it comes, could be cut short. The reason – in order to rein in the deficit, policy makers will have little choice but to raise taxes and cut government spending notwithstanding political protestations to the contrary. Thus one cannot categorically rule out the possibility of the US economy experiencing a double-dip recession.

Wall Street in Denial

What is interesting is that despite a recession that is continuing to inflict widespread damage to the US economy, the consensus view among forecasters is that a recovery is close at hand. In the Wall Street Journal's May survey, the expectation is that the economy will start growing again this summer. According to the survey, real GDP is projected to grow at an annualized rate of 0.6% in the 3rd quarter and the pace is expected to quicken to 2.8% by the 2nd quarter of 2010.

From Recession to Stabilization to Recovery

In the immediate future the US economy still faces significant barriers to recovery and recessionary forces are likely to persist well into 2010. While the worst of the downturn may have passed and GDP may not contract as severely in subsequent quarters, until the credit crunch comes to a definitive end and house prices stop falling (which, in case one needs reminding, triggered the current crisis in the first place), it's unlikely that the US will be able to confidently emerge from this nasty recession.

Nevertheless, there are encouraging signs that the economy is contracting at a decelerating rate and is getting closer to forming a bottom. But before any recovery can take place, the economy has to stabilize, and that outcome is still several months away. Furthermore, once the recovery does get underway, its path will be long and arduous. Those expecting that the economy will bounce back and resume its buoyancy prior to the credit crisis are in for a major disappointment.

The Bottom Line

Faced with the daunting task, at both the consumer and government levels, of restoring their respective balance sheets back to health by de-leveraging and paying down debt, the US economy faces several years of below-trend growth. Unfortunately this means that high levels of unemployment are likely to persist for a prolonged period of time. Not a pretty picture for sure and one that, given the large size of the US economy, has negative implications for world growth as well.

6

The Battle Against Deflation Yet To Be Won

(June 2009)

Amid tentative signs that the recession's grip on the global economy is beginning to ease, investors have started to turn their attention to what they see is the next big problem on the horizon – an outbreak of inflation. The concern here is that the massive increases in government spending and ultra loose monetary policy, aimed at rescuing the world economy from a deep recession, will quickly lead to inflation taking off once the recovery gets underway.

However, a quick look at the facts shows that these concerns over inflation are premature and misplaced.

Firstly, output is still falling in all the major countries and before any meaningful recovery can take place economies need to stabilize. But, with forecasts continually being downgraded – for instance the World Bank now expects global growth to decline by 2.9% this year compared to its March projection of -1.7% – such an outcome keeps getting pushed further back.

Secondly, and more importantly, instead of inflation there is a real danger that the world economy could be facing a period of deflation in the immediate future.

Major Economies Falling into Deflation Territory

Unsurprisingly, with the world economy experiencing its worst slump since the Great Depression of the 1930s, inflation is slowing sharply just about everywhere. For several months now the combination of depressed demand and falling asset prices has been a cautionary red flag that has been pointing in the direction of deflation, or negative inflation.

The latest data shows that the annual rate of inflation has been dropping like a stone and has now turned negative in four of the seven G-7 countries (see Table 1). The consumer-price index in the US fell by 1.3% in the year to May, its sharpest decline since April 1950. During this same period, Britain's annual inflation rate dropped to -1.1%, France's to -0.3% and Japan's to -0.1%.

TABLE 1
G-7 ECONOMIES ENTERING DEFLATIONARY TERRITORY
(Year-over-Year % Change in Consumer Price Index)

	May 2009	2008 Peak
Britain*	-1.1	5.0 Sep
Canada	0.1	3.5 Aug
France	-0.3	3.6 July
Germany	0.0	3.3 July
Italy	0.9	4.1 July
Japan	-0.1	2.3 July
United States	-1.3	5.6 July

* Retail Price Index; Source: National Statistics Agencies

To be sure, the sharp declines in oil and other commodity prices since last summer have been a key reason why inflation has been falling in all G-7 economies. But the question remains, will inflation continue to fall once last summer's run-up in oil prices has dropped out of year-over-year comparisons?

A number of factors including weak global demand, on-going job losses, rising unemployment, stagnant and/or falling wages, and declining household wealth suggest that prices will indeed remain under significant downward pressure for the foreseeable future. Furthermore, despite the deep recession, the real cost of borrowing is rising – the perverse outcome of falling inflation.

Real Cost of Borrowing Rising Rapidly

Over the past several months all the major central banks in the world have been sharply reducing interest rates to combat both the credit crisis and the deepening recession. Moreover, having driven policy rates down to zero or near zero, a number of the central banks including the Federal Reserve and the Bank of England are also pursuing a policy of quantitative easing – effectively printing money – to shore up demand.

However, to date, the impact of monetary policy has yet to make a dent on the real economy. Global growth continues to contract although there are indications that the pace of decline is beginning to slow.

But policy makers now face an even more daunting challenge. With inflation starting to dip into negative territory in several major countries, the effective cost of borrowing is rising. For example, the real rate of interest – which is the difference between the nominal or stated interest rate and the rate of inflation – on 10-year government bonds has risen sharply in all the G-7 economies (see Table 2).To put this in perspective, over the past year central bank policy rates have fallen on average by 263 basis points (2.63%). Nominal yields on 10-year bonds have also declined but only by a paltry 61 basis points.

TABLE 2
REAL RATES RISING SHARPLY
(Yields on 10-Year Government Bonds Less Annual Inflation Rate)

	May 2009	May 2008	Change in Basis Points[1/]
Britain	4.9	0.9	400
Canada	3.3	1.5	180
France	4.2	1.3	290
Germany	3.6	1.0	260
Italy	3.7	1.3	240
Japan	1.6	0.5	110
United States	4.8	-0.1	470

[1/] One basis point is equal to 1/100th of 1%, or 0.01%
Source: See Appendix Table

In inflation adjusted terms, however, real yields on G-7 government bonds have risen on average by 279 basis points during this period. Tellingly, the real cost of borrowing has risen the most in the US and Britain; real rates are up by 470 basis points in the US and by 400 basis points in Britain (see Table 3, column 2).

TABLE 3
Change in Real & Nominal Yields in 10-Year Government Bonds and Central Bank Policy Rates from May 2008 to May 2009

	Real Rate	Nominal Rate	Central Bank Policy Rate
Britain	+400	-130	-450
Canada	+180	-30	-275
France	+290	-70	-300
Germany	+260	-80	-300
Italy	+240	-30	-300
Japan	+110	-30	-40
United States	+470	-60	-175
Average	+279	-61	-263

Sources: National Central Banks, Wall Street Journal Market Data

The trouble is that when the real cost of borrowing rises it discourages both borrowing and spending and acts as a brake on economic activity. Thus, if deflationary pressures were to persist, monetary policy's role in stimulating demand gets effectively shut down as it loses its ability to jump start the economy. After all, central banks can't push nominal interest rates below zero. Once inflation turns negative and deflation takes hold, there is only one way that real interest rates can go, and that is up.

A Japanese Style Deflation Looms

With unemployment rates climbing across the globe and capacity utilization rates falling worldwide, the prospect of inflation taking off anytime soon is simply absurd. Unless growth starts to ramp up smartly in the major economies in the coming quarters, a prospect that is increasingly beginning to look unlikely in an environment where real interest rates are rising, deflation remains a threat to the world economy.

If deflation does get embedded, the world economy could enter a Japanese-style, decade-long deflationary period. Under such a scenario, growth would be meager at best, the unemployment rate would stay high, and the pressure to further reduce incomes and prices would remain relentless.

The Bottom Line

The best anti-dote to deflation is of course strong and sustained economic growth. Unfortunately, given the synchronized nature of this slump and with several key countries saddled with heavy debt loads, the probability of the United States and other major economies experiencing a so-called V shaped recovery – where growth snaps back sharply – is virtually zero. As businesses continue to struggle to work-off excess inventories and rein in supply to match reduced demand, prices will face further downward pressure. As a result, the inflation rate is likely to turn negative in even more countries in the coming months.

Clearly, the battle against deflation has yet to be won and it will continue to dog policy makers in the months to come.

APPENDIX

	10-Year Gov't Bonds Nominal Yields[1]		Consumer Price Index Y/Y% Change		10-Year Gov't Bonds Real Yields[2]	
	May 2009	May 2008	May 2009	May 2008	May 2009	May 2008
Britain	3.8	5.1	-1.1	+4.2	4.9	0.9
Canada	3.4	3.7	+0.1	+2.2	3.3	1.5
France	3.9	4.6	-0.3	+3.3	4.2	1.3
Germany	3.6	4.4	0.0	+3.0	3.6	1.0
Italy	4.6	4.9	+0.9	+3.6	3.7	1.3
Japan	1.5	1.8	-0.1	+1.3	1.6	0.5
USA	3.5	4.1	-1.3	+4.2	4.8	-0.1

[1] Month end figures; [2] Real yield = Nominal yield – inflation
Sources: Wall Street Journal Market Data and National Statistical Agencies

7

This Rally Looks Unsustainable But It's Not All Bad News

July 2009

Markets Rebound Sharply on Signs of Hoped for Recovery

After suffering steep declines over the past several months, stock markets have rebounded sharply since hitting lows for the year in March. In the United States, over the March 9th to June 30th period, the Dow Jones climbed by 29.0%, the S&P 500 by 35.9% and the Nasdaq by 44.6%. Stock markets have also climbed globally with the MSCI world index posting a gain of 40.0% during this period.

The impetus behind these gains has been a growing sense that the deep recession that has gripped the world economy is starting to loosen its hold. Two factors in particular, an easing of the credit crisis and a turnaround in business and consumer sentiment in a number of key countries, are behind this renewed optimism. The decline in the 3-month dollar Libor rate – the interest rate that the banks charge each other for short-term periods – to under 0.6% in June from a high of 4.8% last October is a clear sign that the credit crisis is abating.

On the economy watch front, the OECD's index of composite leading indicators – a system that is designed to provide early signals of turning points in business cycles – provided the backdrop for the rally. After having declined for 22 consecutive months, the index moved up in March and April. Critically though, the readings for both these months have remained below 100 which indicates that businesses are still contracting but the pace of decline is slowing. In other words, the global recession is not over yet but it may be in the process of bottoming.

Investor Enthusiasm Starting to Wane on Mixed Economic Signals

Buoyed by the prospects of an impending recovery, investors have been moving back into equity markets in increasing numbers. But the latest batch of data shows that the economic situation remains fluid. The US employment report for June was a major disappointment as was the unexpected drop in consumer confidence levels.

On the plus side, the latest Institute of Supply Management survey showed that the US service sector, which accounts for over 70% of the economy, is on the cusp of growing again. Moreover, the pace of house price declines has started to slow and the data on pending home sales is edging up which suggests that the 31/2 year old US housing slump may be nearing an end.

Nevertheless, given the conflicting signals about the health of the US economy, investor doubts have inevitably set in again. There is growing concern among some investors that the recent rally in equity markets is not sustainable and there is further downside to come. Many point to the 1929-32 period during the Great Depression where the Dow rallied on several occasions only to fall back again.

In terms of perspective, the current bear market started on October 9, 2007 when the Dow Jones reached an historic high of 14164.53 before falling to 6547.05 on March 9, 2009 for a decline of 53.8%. Indeed, if this date was to be the trough of this bear market, then it would rank as the second worst bear market after the Wall Street Crash of 1929-32 when the Dow plunged by 89.2% (see Table 1).

TABLE 1 MAJOR US BEAR MARKETS (Dow Jones Industrial Average)					
Date	Bear Market	Peak	Trough	Percent Decline	Duration in Weeks
1929-32[1]	Wall Street Crash	381.17	41.22	-89.2%	149
1973-74[2]	OPEC Oil Crisis	1,051.70	577.60	-45.1%	99
2000-02[3]	Dot.Com Crash	11,722.98	7,286.27	-37.8%	143
2007-09[4]	Credit Crisis*	14,164.53	6,547.05	-53.8%	74

[1]/September 3, 1929 – July 8, 1932; [2]/January 11, 1973 – December 6, 1974; [3]/January 14, 2000 – October 9, 2002; [4]/October 9, 2007 – March 9, 2009.
*Advance call on end date for current bear market
Source: Yahoo! Finance

Dow Jones Closely Tracking 1929 – 1932 Market Declines

Comparing the declines of the current bear market to previous major ones does provide some justification for these fears. In terms of trading days, the longest bear market was the Wall Street Crash of 1929-32 which lasted 714 trading days and the shortest one is the current 2007-09 Credit Crisis which has lasted 356 days.

Table 2 shows the percentage loss during the first 365 trading days for each of the four listed bear markets. As you can see, the Dow's plunge of 53.8% in the current bear market is just shy of the 55.1% drop experienced during the first 356 trading days of the Wall Street Crash of 1929-32. Interestingly, the corresponding declines during the 1973-74 and 2000-02 bear markets pale in comparison.

	TABLE 2 MAJOR US BEAR MARKETS (Dow Jones Industrial Average)		
Date	Bear Market	Number of Trading Days	Loss during First 356 Trading Days
1929-32	Wall Street Crash	714	-55.1%
1973-74	OPEC Oil Crisis	482	-18.3%
2000-02	Dot.Com Crash	686	-7.3%
2007-09	Credit Crisis	356	-53.8%

Source: Yahoo! Finance

Repeat of 1929 – 32 Market Collapse – Zero Chance

However, while the Dow may retest the lows of March 9, the probability of it matching the record crash of 1929-32 when the index plunged by 89.2% is remote. And here, a look at the trend in corporate profits during the Great Depression and the current recession provides some important insights. Corporate profits of course decline during recessions but the decline during the Great Depression was unprecedented and catastrophic.

From a high of $10.8 billion in 1929, corporate profits not only nose-dived as the depression unfolded, but turned negative in 1932 and 1933, an event that has never been repeated since. As a percentage of GDP, corporate profits fell from 10.4% in 1929 to a low of minus 0.2% in 1932. During this four-year period, the Dow posted double digit negative returns for each of these years. But, once corporate profits started to recover as the recovery set in, the Dow surged between 1933 and 1935 (see Table 3).

Corporate Profits Turn the Corner in Q1 2009

To be sure, corporate profits have also fallen sharply during the current recession having declined by over $400 billion from its third quarter 2006 peak of $1.7 trillion. In relation to the size of the economy, corporate profits have fallen for nine consecutive quarters tumbling from a high of 12.9% of GDP in Q3 2006 to 8.9% in Q4 2008. However, the good news is that the decline seems to have been stemmed. Corporate profits increased by $48 billion to $1.3 trillion in the first quarter of this year and the ratio of corporate profits-to-GDP ratio rose to 9.3%.

TABLE 3
US CORPORATE PROFITS, NOMINAL GDP, AND THE DOW JONES
(1929 – 1935)

Year	Corp Profits ($ billions)	Nominal GDP ($ billions)	Corp Profits/ GDP Ratio (%)	Dow Jones
1929	10.8	103.6	10.4	-17.2%
1930	7.5	91.2	8.2	-33.8%
1931	2.9	76.5	3.8	-52.7%
1932	-0.2	58.7	-0.3	-22.6%
1933	-0.1	56.4	-0.2	+63.7%
1934	2.5	66.0	3.8	+5.4%
1935	4.0	73.3	5.5	+38.5%

Source: Bureau of Economic Analysis, Yahoo! Finance

Nevertheless, these are still early days and until the recession finally releases its tight grip on economic activity there could well be more downward pressure on corporate profits. While the recent rally may fade, we are nevertheless a long way from experiencing anything like the total meltdown in corporate profits that occurred during the

Great Depression. When all is said and done, it is corporate profits that underpin the long term direction of stock markets.

The Bottom Line

What distinguishes the 1930s from today is that during the Great Depression there was a total collapse in corporate profits which completely crushed the stock market. To match the Wall Street Crash of 1929-32 the Dow would have to fall all the way to 1,530 from its record high of 14,164.53 set back in October 2007. But such an outcome is highly unlikely today. After all, US businesses still managed to generate total profits of $1.3 trillion in the first quarter. Moreover, if the first quarter up-tick in profits can be sustained in the coming months, then the recent gains in the stock market may indeed be signaling that this latest bull market is now underway.

Only time will tell if March 9, 2009 – the start of the latest rally – represents a true inflection point. But, if history is any guide, it is the trend in corporate profits in the coming quarters that will determine the market's direction. As always, investing in the stock market will continue to challenge investors as it has since its creation eons ago.

Only a dupe would expect it to go up in a straight line.

8

After The Recession
An Erratic & Jobless Recovery

(August 2009)

In the United States – the world's dominant economy – there are signs that the recession is coming to an end. After shrinking at an annual rate of 6.4% in the first three months of this year, real GDP fell by 1% in the second quarter, a clear sign that the pace of contraction is slowing. Moreover, there are indications that economic activity is starting to revive, albeit from depressed levels. The housing market is stabilizing, new home sales have ticked up, the manufacturing sector is poised for expansion, and the pace of job losses is diminishing.

With the focus of investors now turning to the upcoming recovery, the two key questions they will have to address are – how strong will the recovery be and, perhaps more importantly, how sustainable. To gain some insight into these issues, a look at a number of past US recessions and their subsequent recoveries may be helpful in providing clues as to the future direction of the American economy.

Impact of US Recessions

As you can see from Table 1, there have been a total of six recessions of varying depths and durations in the US since the 1970s. The current recession, which started in December 2007, is now in its 19th month and is the longest one since the Great Depression of the 1930s. Moreover, its impact on output and employment has been nothing short of devastating. The 3.7% decline in GDP and the horrendous loss of 6.7 million jobs attest to the severity of this downturn.

By way of comparison, with the exception of the 2001 recession where output actually increased, real GDP declines have ranged from -3.2% in the 1973-75 slump to -1.4% in the 1990 recession. On the employment front, job losses in the other five recessions have ranged from about 1 million in 1980 to more than 2.8 million in 1981-82.

TABLE 1
IMPACT OF US RECESSIONS ON OUTPUT AND EMPLOYMENT

Recession Periods	Duration (Months)	Percent Decline in Real GDP	Total Job Loss (000s)	% Job Loss from Peak
Dec 2007(Q4) to Jul 2009(Q2)[1]	19	-3.7%	-6,664	-4.8%
Mar 2001(Q1) to Nov 2001(Q4)[2]	8	+0.7%	-1,599	-1.2%
Jul 1990(Q3) to Mar 1991(Q1)	8	-1.4%	-1,240	-1.1%
Jul 1981(Q3) to Nov 1982(Q4)	16	-2.6%	-2,824	-3.1%
Jan 1980(Q1) to Jul 1980(Q3)	6	-2.2%	-968	-1.1%
Nov 1973(Q4) to Mar 1975(Q1)	16	-3.2%	-1,260	-1.6%

[1] assuming recession trough
[2] In the 2001 downturn, the US economy did not contract for two consecutive quarters, the standard rule of thumb used to delineate recessions. Instead, GDP declined in the first quarter and again in the third quarter of 2001

Source: NBER, Bureau of Economic Analysis, Bureau of Labor Statistics

The Path to Recovery

Table 2 shows the growth in real GDP and employment in the first year of recovery following each of the past five recessions since the

1970s. As you can see, a number of key points leap out from the table.

Firstly, the rebound in growth following the end of the 1981-82 and the 1973-75 recessions, both of which lasted 16 months, was robust as GDP surged by 7.7% and 6.2% respectively. Output also recovered quickly in 1980's short recession which lasted only 6 months. In contrast, the recovery following the recessions of 1990-91 and 2001, each of which lasted 8 months, was much more muted. Real GDP increased by 2.6% and 1.9% respectively in the year following these recessions.

TABLE 2 PATH TO RECOVERY Growth in Output & Employment in 1st Year of US Recoveries			
Recession Trough	Real GDP	Employment pct. change	Employment (000s)
2001 Q4	1.9%	-0.4%	-562
1991 Q1	2.6%	-0.2%	-239
1982 Q4	7.7%	+3.5%	+3,084
1980 Q3	4.4%	+2.0%	+1,762
1975 Q1	6.2%	+3.1%	+2,400

Source: NBER, Bureau of Economic Analysis, Bureau of Labor Statistics

Secondly, although GDP increased in the first year of recovery following each recession, the same was not the case for employment. Powered by the strong resurgence in output growth, the rebound in jobs after the recessions ended in the 1980s and 1970s was strong. For example, in the year following the recession trough in the fourth quarter of 1982, the US economy generated 3.1 million jobs; more than recouping the 2.8 million jobs that were lost during the recession. Similarly, in the year following both the 1973-75 and 1980 recessions, the number of jobs that were created (2.4 million and 1.76 million respectively) more than made up for the previous employment losses.

In sharp contrast, however, jobs continued to be eliminated in the year following the 1991 and 2001 recessions. The reason: the pick-up in aggregate demand was simply not strong enough to spur job creation. After suffering employment losses of 1.6 million and 1.2 million during the 8-month recessions of 2001 and 1991, a further 562,000 and 239,000 jobs were eliminated during the recovery year.

In analyzing the path to recovery following previous US recessions, the message is clear – it takes a strong resurgence in demand, with the economy expanding at well-above trend levels, for job creation to return. Failing that, the economy is saddled with a jobless-recovery. This, in turn, brings us to the current outlook for the US economy. And here, unfortunately, the prospects don't look good.

The US economy is being weighed down by a myriad of imbalances and obstacles including soaring budget deficits, high unemployment levels, sharp declines in household wealth, and by consumers – a mainstay of the economy – who are reducing spending, intent on paying down debt, and saving more. Taken together, these are all powerful factors that stand in the way of the US economy experiencing a full and sustained recovery.

After this Recession – A Jobless Recovery

Assuming that the second quarter of 2009 does indeed turn out to be the end of this recession, this time around, unlike the strong rebound in GDP growth following the 1973-75 and 1981-82 recessions, there will be no quick return to healthy growth. Instead, the recovery will be slow and halting; economic growth is forecast to average only between 1.5% and 2% in the first year of recovery. As a consequence, it likely won't be until 2011 at the earliest before output regains its previous peak.

Similarly, it will take a number of years before total employment surpasses its pre-recession level. Unfortunately, given the projected tepid pace of growth, job losses will continue to mount. As you can glean from Table 2 there is, not surprisingly, a strong statistical relationship between growth in GDP and employment. Simply put – the higher the rate of growth in output, the higher the growth in employment and vice-versa. This close correlation between output and employment allows one to get a handle on the likely impact on jobs under different growth scenarios. For example, if the US economy were to grow by 1.5% during the recovery year, this could result in a further 850,000 jobs being eliminated; a 2% increase in GDP would still lead to a loss of close to 375,000 jobs.

To put this in perspective, from the start of the recession in December 2007 to July 2009, the US economy lost an average 350,000 jobs a month. In the first year of this recovery, an additional 30,000 to 70,000 jobs per month could be lost. There is little question that this recovery will prove to be the exception to the rule and will lay to rest the oft-quoted mantra that the 'deeper the recession, the faster the recovery'.

The Bottom Line

While the worst may be over for the US economy, the chances of it experiencing a rapid and lasting economic recovery are remote. That's not to say that a sharp bounce in growth in the near-term is impossible. The US economy may indeed experience a so-called V-shaped recovery over the next quarter or two as exports, government spending and an anticipated boost from inventory rebuilding shore up growth.

But don't be fooled. While it may beguile market participants into believing that the 'good times' are returning again, the reality is that the American economy faces a long and tough slog ahead.

9

What's Driving The Surge In Global Equity Markets?

(September 2009)

Despite a decidedly weak global macroeconomic environment, with countries struggling mightily to pull themselves out of the worst economic downturn since the Great Depression, stock markets have been rallying strongly around the world since early March. What is amazing is that all equity markets in both developed and emerging economies have posted double-digit returns. Between March 9 and August 31 of this year, stock market increases in the G-7 countries have ranged from a low of 38.6% in Britain to a high of 77.6% in Italy (see Table 1).

Similarly, stock prices in emerging markets have also surged during this period with India's market gain of 92.0% leading the pack.

But what is truly astounding is that stock markets have also rocketed-up even in the world's hardest hit economies. A number of countries, including the Baltic States of Latvia, Estonia and Lithuania as well as Turkey and the Ukraine, are literally in free-fall (Table 2).

TABLE 1
GLOBAL STOCK MARKET PERFORMANCE
(in local currency)

	Index Aug 31 2009	Since Mar 9 2009 Low	Previous High	From High to Aug 31 2009
G7 ECONOMIES				
USA (DJIA)	9,496.3	45.0%	Oct 9, 2007	-33.0%
Canada (S&P/TSX)	10,868.2	43.6%	Jun 18, 2008	-27.9%
Japan (Nikkei 225)	10,492.5	48.1%	Jul 9, 2007	-42.5%
Britain (FTSE)	4,908.9	38.6%	Jun 15, 2007	-27.1%
France (CAC 40)	3,653.5	45.0%	Jun 1, 2007	-40.8%
Germany (DAX)	5,464.6	48.0%	Jul 16, 2007	-32.6%
Italy ((MIB)	22,420.4	77.6%	May 18, 2007	-49.5%
EMERGING MARKETS				
Brazil (BVSP)	56,489.0	53.7%	May 20, 2008	-23.2%
Mexico (IPC)	28,129.9	65.8%	Oct 18, 2007	-14.3%
Venezuela (IBC)	50,524.1	34.7%	Jan 8, 2007	-18.5%
Russia (RTS)	1,066.5	68.0%	May 19, 2008	-57.1%
China (SSEC)	2,667.8	25.9%	Oct 16, 2007	-56.2%
India (BSE)	15,666.6	92.0%	Jan 8, 2008	-24.9%
South Africa (JSE)	24,929.4	37.3%	May 22, 2008	-25.0%

Source: Bloomberg

TABLE 2
UP, UP, AND AWAY!
Stock Market Returns of World's Hardest-Hit Economies

	Index Aug 31 2009	Since Mar 9 2009 Low	Year-over-Year Pct. Change in GDP Q2 2009*
Lithuania (OMXV)	277.9	84.2%	-20.2%
Latvia (OMXR)	312.6	53.9%	-18.7%
Ukraine (PFTS)	469.1	133.7%	-18.0%
Estonia (OMXT)	399.2	62.9%	-16.1%
Turkey (ISE 100)	46,551.2	101.9%	-13.8%

* except for Turkey Q1 2009. Source: Bloomberg, Eurostat, IMF

GDP growth in these economies has fallen precipitously over the past year plunging them well past recession territory and into a Great Depression zone. Yet, their stock markets have all posted double-digit returns. Indeed, stock prices in both the Ukraine and Turkey have more than doubled over the past six months.

What's Driving the Rally?

This begs the question: what's driving the global rally and, more importantly, is it sustainable on a going forward basis?

To be sure, stock markets will, and do, gyrate in the short-term for reasons that are based more on rumours and sentiment. But what distinguishes the current period is that stock markets have risen sharply in every single country on the planet despite the still dismal economic picture. Nevertheless, there are a couple of key reasons why markets have rallied.

Firstly, faced with falling revenues, companies have been relentlessly cutting costs and laying off workers. It is these measures that have helped to prop-up earnings and boosted the share prices of many companies. The trouble is that cost-cutting is a one-shot deal and by itself it cannot fuel the stock market indefinitely. For that to happen, it will take a sustained pick-up in demand and 'real' growth in profits for markets to confidently move up.

Secondly, the massive amount of liquidity that central banks have pumped into the system is not finding its way into productive investments. But this is hardly surprising given that capacity utilization rates are at record lows in several countries. With demand falling, businesses have little interest in expanding capacity and building new factories. Until global demand picks up on its own volition, that is, it is not dependent on a continuing string of government bailout packages, the recovery will remain both halting and weak. However, in the meantime, in the hunt for better returns, much of that money has been going into equity markets sending asset prices back into bubble territory.

Equity Markets Still Below Record Levels

Nevertheless, in spite of the recent surge, the world's equity markets still have a long way to go before they recoup all of the losses incurred since hitting their respective previous peak levels. For example, although the Dow Jones Industrial Average is 45.0% above its March 9 low, it is still well below its record high reached in October 2007. To put this in perspective, the index will need to rise by another 4,668 points (or 33%) from its August close to get back to its October 9, 2007 high of 14,164.5.

Similarly, on the emerging markets front, the stock market indexes in China and Russia are still at less than half their peak levels. Stocks in China will need to rise by 128.4% in order for the market to return to its previous high and in the case of Russia it will take an increase of 133.3% for the RTS index to get back to where it was in May of 2008.

A Red Flag – Are Markets Poised to Crash Again?

The strong rebound in global equity markets in the past six months is a red flag and suggests that markets have divorced themselves from reality by ignoring the fundamentals. While the pace of economic contraction has clearly declined and a cyclical trough is in sight, recent media headlines suggesting that a global recovery is underway may well turn out to be premature. Indeed, there is a real risk that once the recession ends the ensuing recovery's momentum may fizzle out when the stimulus effect from expansionary economic policies runs out.

The bottom line is that there is no quick fix to the global economy. Although a few countries including Germany, France, and Japan are starting to pull out of the recession, the outlook is shrouded in uncertainty. Demand remains weak as heavily indebted consumers retreat back into their shells, businesses are cutting back on their planned investment spending, the job market remains bleak, and unemployment levels continue to climb. This hardly points to a picture of burgeoning growth. On the contrary, growth could just as easily turn negative again in the next quarter or two. This certainly is a worry for investors and there is growing concern that stock markets may well be poised to crash again before long.

Could History be Repeating Itself Again?

While nobody has a crystal ball that can predict the markets future direction, it's worth recalling that during the Great Stock Market Crash of 1929-32, the Dow Jones Industrial Average rallied on a number of occasions only to fall back down again. During the thirty-four month bear market, the Dow rallied on five separate occasions posting an average increase of 31.7%. But, during the markets down phase following each of these rallies, the Dow declined on average by 44.8%, a much larger fall that more than swamped the rallies.

When it was finally all over and the Dow had hit bottom, the index had lost a gut-wrenching 89.2% of its value. In money terms, a $10,000 investment made at the stock market's peak on September 3, 1929 was worth only $1,080 on July 8, 1932 when the rout ended.

The Bottom Line

This is clearly a market for professional traders. Those who are fleet of foot and can get out before the market slumps and get back in again when it makes one of its periodic runs will have the upper-hand. It goes without saying that this is easier said than done. But, while the traders in investment houses have their work cut out for them, this is definitely not a market-friendly environment for the average investor. In these uncertain times, a buy & hold strategy is likely to be severely tested.

Consider the case of Japan which provides a cautionary tale. During the boom years from 1984 to 1989, when growth in real GDP averaged 4.5% a year, the Nikkei 225 stock index posted an average annual compound return of 25%. After hitting a peak on December 29, 1989 when the index closed at 38,916, the market crashed. Unable to shake its economic woes despite numerous stimulus measures, Japan's economy was mired in a prolonged on-again, off-again type of economic recovery from 1990 to 2008. During this 19-year period, when growth in GDP averaged only 1.5%, the Nikkei suffered a cumulative loss of 77%, for an average annual decline of *minus* 7.5%.

Investors who blindly buy into the popular view that 'stock markets are forward looking' are likely to be disappointed. Tread Cautiously!

10

Why A Weaker Canadian Dollar Matters

(October 2009)

Not to be outdone by the surge in equity markets, the Canadian dollar has also been on a tear since early spring. Boosted in part by the pick-up in commodity prices that have been rising in anticipation of a global recovery as well as the negative outlook for the US dollar, the loonie has been a key currency of choice among traders. With demand for the currency rising, the exchange rate has appreciated by about 19% going from 79 cents (US) in March to over 94 cents (US) in early October.

While the rise in stock markets is welcome news and has helped investors to recoup some of their earlier losses, the same cannot be said for the Canadian dollar. For an export dependent economy that is struggling to unshackle itself from the grip of recession, a strengthening loonie is the last thing that the country needs. Indeed, given the severity of the global downturn and the resulting collapse in international trade, Canada's export and manufacturing sectors have been hit particularly hard.

The Gutting of Exports and Manufacturing

Alarmingly, after peaking in the second quarter of 2007, exports have now declined for an unprecedented 8 consecutive quarters. In volume terms, Canada now exports about 25% less than it did two years ago and, relative to the nation's GDP, the share of exports has fallen from 39% to 31%. Moreover, in line with the deterioration in trade, the country recorded a deficit in its current account in the second quarter – it's first since 1976.

The impact on manufacturing has been equally devastating; the sector has lost 301,000 jobs in the two-year period to August 2009. To be sure, given rising productivity levels, greater efficiency, and the emergence of China as a manufacturing powerhouse, manufacturing's share of total employment has been steadily declining in all western economies for several years now. Nevertheless, in Canada's case the sharp appreciation of the loonie since hitting a record low of 62 cents (US) back in January 2002 has not helped matters.

The Canadian Dollar During Previous Recoveries

The recent appreciation of the Canadian dollar clearly presents a major obstacle to lifting the economy out of this recession. What is interesting to note is that in previous recessions, it was a depreciating dollar that played an important role in rescuing the economy. For example, in the two years following the 1990-91 and 1981-82 recessions, the dollar's depreciation boosted the competitiveness of the export sector and provided an extra lift to the recovery. The dollar fell from 86.7 cents (US) following the end of the recession in April 1991 to 79.2 cents (US) in April 1993. Similarly, during the recovery from the 1981-82 recession the dollar went from 81.3 cents (US) in October 1982 to 75.9 cents (US) in October 1984.

Table 1 highlights the economy's recovery path in terms of total output (GDP) and exports. In the first year of recovery following the

1981-82 recession, GDP increased by 6.3% but exports surged by 19.6%. Similarly, in the second year of the recovery, GDP increased by 5.7% but exports rose by 13%. While the pickup in overall growth was much more muted in the two years following the 1990-91 recession, without the contribution from the export sector, the economy would have likely stayed in recession for longer.

TABLE 1
PATH TO RECOVERY
Growth in Real GDP & Exports

Recession Trough	1st Year of Recovery		2nd Year of Recovery	
	GDP	Exports	GDP	Exports
1982 Q4	6.3%	19.6%	5.7%	13.0%
1991 Q1	0.8%	8.7%	1.8%	11.3%

Source: Statistics Canada

What the Bank of Canada can do to stem the Rise

While the recent surge in the loonie is clearly not good news, there is little that the Bank of Canada can do to ameliorate the situation. Interest rates are already at rock bottom (the benchmark rate is at 0.25%) and the central bank can't lower them further to push the dollar down. In recent weeks the central bank has tried to 'jawbone' or talk down the currency. However, despite these warnings, the loonie has continued to appreciate as currency strategists remain bullish on Canada's prospects relative to other major economies.

The Bank of Canada has suggested that, if need be, it will use quantitative easing to reign in the currency. To be sure, buying up government bonds will push up bond prices and lower yields, but it is highly unlikely that such a strategy will work in lowering the Canadian dollar given that there is little difference between interest rates in the two countries. For example, at the end of September, the difference between the yields on 3-month treasury bills was only 6 basis points (0.06%) and the spread between US and Canadian 10-year government bonds was only 1 basis point (0.01%).

A weaker Canadian dollar would clearly be helpful for the recovery and would provide a much-needed tonic for the economy. A lower dollar would boost manufacturing production, lift exports and employment and keep the Canadian economy moving forward. On the flip side however, a weaker dollar will lead to higher import prices and put upward pressure on inflation. But the link between a lower dollar and inflation should be of little concern at this stage of the cycle. Indeed, with the annual inflation rate now in negative territory – the consumer price index fell 0.8% in August from a year earlier – a lower dollar is precisely what the economy needs to ensure that deflationary forces don't become entrenched.

The Bottom Line

Unfortunately, as noted above, there is little that the Bank of Canada can do to stem, let alone reverse, the rise in the loonie. With the benchmark rate already at a record low, the bank has very little wiggle room and effectively finds itself boxed in. A high dollar will continue to put a choke-hold on the export sector and will be a major impediment to the recovery.

Of course, unless the US economy rebounds strongly over the coming months, a prospect that is increasingly looking unlikely, even a weaker currency will be of little help. In a worst case scenario, the combination of a faltering US economy and a high Canadian dollar would effectively scuttle Canada's recovery.

Here's hoping that such a 'double-dip' recession scenario for the Canadian economy doesn't come to pass.

11

World Economy Slipping Into Deflation Territory

(November 2009)

The world economy is now heading towards deflation and the latest data shows that the annual rate of inflation has not only slowed sharply over the past several months but it has started to tumble into negative territory in dozens of countries (see Table 1). This global multi-country fall in consumer-price inflation is quite unprecedented and attests to the severity of the ongoing global recession.

The inflation rate in the United States fell to –1.3% in September of this year, down from 4.9% a year earlier and prices in the US have now been in negative territory for seven consecutive months. Similarly, inflation was negative in the other G7 economies of Britain, Canada, France, and Germany. Unhappily, Japan, after finally managing to finally extricate itself from a decade-long bout of falling prices, finds itself mired yet again in deflation with prices now falling at an annual rate of –2.2%.

Inflation is also turning negative in many of the emerging economies. In China, the world's third largest economy, inflation was –0.8% in September and the index has now been in negative territory for eight consecutive months. Similarly, prices are falling in the Asian economies of Malaysia, Singapore, Taiwan and Thailand and deflationary forces are also plaguing a number of the emerging Eastern European economies including Estonia, Slovenia and the Czech Republic. In Latin America, Chile's annual inflation rate has nose-dived from 9.2% a year ago to –1.1% this September.

TABLE 1
DEFLATION TAKES HOLD IN THE GLOBAL ECONOMY
(Year-over-Year % Change in Consumer Prices)

Country	Sept 2008	Sept 2009	Country	Sept 2008	Sept 2009
USA	+4.9	-1.3	Cyprus	+5.0	-1.2
Japan	+2.1	-2.2	Czech Rep	+6.4	-0.3
China	+4.6	-0.8	Estonia	+10.8	-1.7
Germany	+3.0	-0.5	Finland	+4.7	-1.0
France	+3.4	-0.4	Slovenia	+5.5	-0.1
Britain	+5.0	-1.4	Switzerland	+2.8	-1.1
Canada	+3.4	-0.9	Sweden	+4.4	-1.6
Spain	+4.6	-1.0	Chile	+9.2	-1.1
Belgium	+5.5	-1.0	Malaysia	+8.2	-2.0
Ireland	+3.2	-3.0	Singapore	+6.7	-0.4
Portugal	+3.2	-1.8	Taiwan	+3.1	-0.9
Luxembourg	+4.8	-0.4	Thailand	+5.1	-1.0

Sources: Eurostat, The Economist, National statistics offices

Stimulus Measures Being Thwarted

Since the onset of the global recession in late 2007 policy makers have tried their utmost to spur demand and stop the economy from being sucked into a deflationary spiral. On the monetary policy front, the world's major central banks have slashed interest rates to record lows and a number, including the Federal Reserve and the

Bank of England, have embarked on an aggressive campaign of printing money via 'quantitative easing'.

Similarly, on the fiscal policy front, governments have jacked up spending and poured trillions of dollars into their respective economies in order to shore up demand. Taxes have also been cut.

Yet, despite these aggressive policy measures, the world economy remains in a deep funk and consumer prices at the aggregate level have continued to decline.

Deflation's Destructiveness

While a mild, and temporary, bout of deflation is no cause for alarm, it is another story if deflationary forces were to become entrenched. The longer it lasts the more grievous damage it inflicts on the economy. Deflation has a pernicious effect on a number of key economic variables. It reduces consumer spending, it increases the real value of debt, it raises the cost of borrowing, and it renders monetary policy impotent.

Initially, consumers welcome cheaper prices, but once deflation takes hold and consumers expect prices to fall further, they delay making purchases. This delay leads to a fall in consumer spending which in turn leads to lower economic growth. To clear their shelves, businesses respond by further lowering prices and in order to bring costs into line with reduced demand, start laying-off workers.

Moreover, deflation increases the real value of debt. As prices fall and wages come under downward pressure, the burden of debt rises. The level of debt stays the same but servicing it becomes more onerous as incomes drop.

By raising the real cost of borrowing, deflation also discourages both borrowing and spending. For example, if interest rates are, say, at 4% but inflation is at *minus* 2%, the real cost of borrowing is 6%. Indeed, loans to households and companies are growing at their slowest pace in decades in most OECD economies. Not surprisingly, given the depth of the global downturn and the aftermath of the credit crunch, banks are focusing on repairing their balance sheets and are reluctant to lend.

On the other hand, faced with mounting job losses and growing economic uncertainty, households are retrenching and paying down debt. Given these circumstances, there is little appetite for taking on additional debt. Similarly, with capacity utilization rates hovering at record lows in all the major economies, the need for companies to borrow funds in order expand plant and equipment is simply not there.

A Self-Sustaining Recovery is Still a Long Way Off

While there are welcome signs that the worst of the global recession is now behind us – GDP growth has turned positive in a number of the major economies including the United States, Germany and France – the foundation of the economic recovery is neither stable nor solid. There remains a very real concern about the sustainability of the recovery which is largely being driven by the various government stimulus measures and other temporary factors. But in light of the fact that job losses continue to mount, consumer confidence levels remain shaky, credit conditions remain tight, and private sector loan demand is virtually non-existent, it would seem that a self-sustaining recovery is still a long way off.

The key question remains – are the flickering signs of improvement in the global economy a precursor to sustainable growth or does it merely reflect a pause before the economy starts to downshift again? Much will depend on the sources of growth and its trajectory over the next few quarters. Crucially, until the private sector, especially in western economies, starts to ramp up production and resumes hiring again, the recovery could well falter.

Moreover, the danger is that if growth fails to revive once the stimulus ends, the world may find itself back at square one once again. But, this time it will be saddled with a humongous amount of government debt.

The Bottom Line

In the concluding sentence of the June 2009 edition of the monthly commentary I cautioned that '….the battle against deflation has yet to be won and it will continue to dog policy makers for months to come'. With deflation now starting to take root in dozens of countries the matter is even more urgent.

12

What's In Store For 2010?
— *Some Observations* —

(December 2009)

No Rapid Rebound from the Great Recession

Recent GDP data shows that economic growth, after falling precipitously for three or four quarters, has resumed in a number of the major economies including the United States, Canada, France, Germany, and Japan. As these economies have started to exit from the global recession attention has now turned towards assessing the underlying strength and durability of the recovery.

To be sure, aggressive monetary policy easing and the enactment of massive government stimulus measures have – finally – succeeded in turning around the economy. But on a going forward basis, as the stimulus winds down, the sustainability of the recovery will increasingly depend on a pick-up in household and corporate spending. The bottom line is that unless private sector demand kicks in, the recovery will be shaky. And here, unfortunately, the prognosis for 2010 does not look particularly good.

With consumers intent on paying down debt and corporate loan demand either flat or turning negative in several of the big developed economies, the de-leveraging process still has a long way to go. As a consequence, aggregate demand is expected to remain quite anemic. What this means is that an on-again/off-again recovery will likely plague the G-7 economies for a number of quarters. Policy makers who are expecting, or indeed banking, on a rapid and sustained rebound from the Great Recession are therefore likely to be disappointed.

Charity Begins at Home

Unless economic growth picks up smartly in the G-7 economies, the odds of which as noted above appear to be low, job losses will unfortunately continue to mount well into 2010. Confronted with rising unemployment and growing poverty levels the pressure will be on for governments to act and 'do something'. And here, politicians will face an uphill battle in resisting calls for protectionism as a growing segment of the electorate will increasingly demand that 'charity begins at home'. Already there are signs that governments are yielding to this call.

According to the non-partisan Global Trade Alert's second report entitled 'Broken Promises: A G20 Summit Report' it found that between November 2008 and September 2009 "…121 beggar-thy-neighbor measures have been implemented by G20 governments since last November. Every three days a G20 government has broken their no-protectionist pledge".

For governments, already strapped for cash and with little room to maneuver on the economic policy front, 2010 promises to be every bit as tough and challenging as 2009. Maybe even more so. As the pressure mounts and populist anger begins to rise, we may be entering an era where politics trumps economics and governments start to put up the shutters to open markets and open trade. But, history shows that while protectionist measures may well serve as a short-term palliative, its effects are very deleterious in the longer run resulting in sub-par growth and diminished standards of living.

Global Stock Markets will be in Corrective Mode

Notwithstanding the severe fall-out in output and employment from the Great Recession, one of the hall-marks of 2009 is that, astonishingly, virtually every single stock market in the world posted double-digit returns. In the advanced economies, gains ranged from a low of 13.1% (to December 4) for Japan's Nikkei 225 Index to a high of 29% for Australia's All Ordinaries index. On the emerging markets front, equity markets in Indonesia, China, and India topped the leader board posting returns of 85.3%, 82.2%, and 77.3% respectively.

This naturally begs the question: can markets continue to climb after this spectacular run-up in stock prices or are they set to correct? While nobody can foretell the future, much will depend on the trajectory for economic growth and the stance of economic policy in the months ahead. Given that governments in North America and several European economies have indicated their intention to maintain the various fiscal stimulus measures through 2010 this should be a positive for equities, at least for the first half of next year.

On the monetary policy front, interest rates also are expected to remain low well into 2010 which will further lend support to the equity markets. Moreover, because there is so much spare capacity available, worries about an imminent break-out of inflation is clearly unwarranted. Indeed, if anything, should overall demand remain depressed there is a greater risk of deflation taking hold in the near term.

But that doesn't mean to say markets won't experience significant stress at times. Given the ongoing uncertainty about the robustness of the global recovery, I fully expect that markets will be punctuated with (mini and/or major) corrections along the way as it strives to move higher.

More Dubai Style Debt Bombs Likely

The recent announcement by Dubai World – the investment arm of the United Arab Emirates – that it was seeking to delay payments on about $60 billion in debt shook global markets and is a wake-up call that there could be more such debt bombs lurking in the global financial landscape. As governments and companies in all regions of the world find themselves laden down with mountains of debt, Dubai World's decision to restructure its debt and delay making payments has sharply focused the minds of investors.

The key concern here is that while it is generally accepted that central banks and governments will come to the rescue of distressed commercial banks, albeit with stringent conditions attached, there is no assurance that such support will also automatically be extended to heavily indebted companies that run into financial difficulties. Indeed, the UAE's refusal to guarantee the debts of Dubai World is precedent setting and is forcing investors to re-assess the risks of holding corporate debt particularly from emerging markets. The risk of contagion of even a few corporations defaulting on their debt is the Achilles heel that could well roil markets in 2010.

PART 2

The Year 2010

13

Global Stock Market Performance In The 2000s
— *Winners & Losers* —

(January 2010)

Last year ended on a stellar note for the world's stock markets. In anticipation of a global recovery from the Great Recession, stock prices soared in all the regions of the world. In the advanced economies, the MSCI index for developed markets posted a gain of 30.8%. While impressive, these gains were no match for the performance of the emerging markets as the MSCI index for those markets rocketed up by 78.7%.

As we enter the second decade of the 21st century, it is both interesting and instructive to take note of how equity markets performed during the first decade of the new century. And here, what jumps out immediately is that, at the aggregate level, this decade also belonged to the emerging economies. Over the 10-year period from December 1999 to December 2009, the MSCI emerging markets index delivered an average annual return of 10.1%, far outstripping the returns from the developed markets which managed to eke out a niggardly annual return of 0.2%.

In money terms, an initial investment of $10,000 on December 31, 1999 in the stock markets of the developed world would have grown to $10,200 by the end of 2009, for a paltry gain of $200. In sharp contrast, a similar investment in the emerging markets would have netted an investor a gain of close to $16,200, or 81 times more!

Gauging G-20 Stock Markets

Of course, as we all know, averages tend to mask more than they reveal and inevitably raise more questions. In particular, for investors, the answers to the following two questions should be especially illuminating.

Firstly, did all emerging markets register stellar gains in the first decade of the 21st century and,

Secondly, were all the stock markets of the developed economies in the dumpster?

Table 1 highlights the stock market performance of each of the G-20 countries during the past decade. This group, which accounts for about 80 percent of global output and trade, is made up of 19 developed and developing countries and also includes the European Union. Advanced economies in the G-20 are Australia, Canada, France, Germany, Italy, Japan, South Korea, the United Kingdom, and the United States. Emerging economy members are Argentina, Brazil, China, India, Indonesia, Mexico, Russia, Saudi Arabia, South Africa, and Turkey

Stock Markets in Several Advanced Economies Inflict Sharp Losses

As one can see the performance of stock markets in the advanced economies in the last decade was a mixed bag with only three posting gains and six delivering losses. In this group, South Korea's stock market was the best performer. In nominal terms, it increased at an annual rate of 5.1%, for a total gain of 65% over the ten-year period.

TABLE 1
G-20 STOCK MARKET PERFORMANCE IN THE 2000s
DECEMBER 1999 – DECEMBER 2009
(IN LOCAL CURRENCY TERMS)

		Nominal Returns (%)		Inflation (CPI)	Real Returns[1] (%)	
		Ann[2]	Cum[3]	Ann[2]	Ann[2]	Cum[3]
		(1)	(2)	(3)	(4)	(5)
Advanced G-20 Economies						
South Korea	Kospi	5.1	64.5	3.1	2.0	21.9
Australia	Ordinaries	4.5	55.3	3.2	1.3	13.8
Canada	S&P/TSX	3.4	39.7	2.1	1.3	13.8
Germany	DAX	-1.5	-14.0	1.6	-3.1	-27.0
Britain	FTSE 100	-2.4	-21.6	1.8	-4.2	-34.9
United States	S&P 500	-2.7	-23.9	2.6	-5.3	-42.0
France	CAC 40	-4.1	-34.2	1.9	-6.0	-46.1
Japan	Nikkei 225	-5.7	-44.4	-0.2	-5.5	-43.2
Italy	MIB	-5.9	-45.6	2.3	-8.2	-57.5
Emerging G-20 Economies						
Russia	Micex	24.6	802.0	14.1	10.5	171.4
Mexico	Bolsa	16.2	348.8	5.2	11.0	183.9
Argentina	Merval	15.8	333.6	8.5	7.3	102.3
Brazil	Bovespa	14.9	301.1	6.9	8.0	115.9
Indonesia	JKSE	14.1	274.0	8.4	5.7	74.1
India	Sensex 30	13.3	248.6	5.4	7.5	106.1
Turkey	ISE 100	13.3	248.6	23.1	-9.8	-64.4
South Africa	JSE	12.5	224.7	6.1	6.4	87.0
Saudi Arabia	TASI	11.7	202.4	2.0	9.7	152.4
China	Shanghai	8.8	132.4	1.9	6.9	94.9

[1] Real Return = Nominal Return minus Inflation; [2] Annual Rate of Change; [3] Cumulative Returns; Sources: Bloomberg, IMF, Country stock exchanges

The stock markets in the resource rich economies of Australia and Canada also rewarded investors posting annual gains of 4.5% and 3.4% respectively. But investors in the stock markets of the United States, Europe, and Japan all suffered significant losses. Italy's market was the worst performer of the decade posting an annual decline of 5.9%, for a total loss of 45.6% (see Table 1, columns 1 & 2).

In marked contrast, the stock exchanges in all the major emerging markets not only posted positive returns in the last decade but also outstripped, by a wide margin, the market performance of the advanced economies. When measured in nominal terms, Russia's market, with an annual gain of 24.6%, topped the leader board and won the stock market sweepstakes. An investment of 10,000 rubles at the end of December 1999 was worth 90,200 rubles at the end of 2009, for a cumulative gain of 802%! Interestingly, among the emerging markets group, China's market was, comparatively speaking, the worst performer posting a cumulative gain of 132%, or 8.8% a year.

Adjusting Stock Market Returns for Inflation

Although many investors generally tend to ignore the impact of inflation when assessing stock market returns, incorporating price changes gives one a more accurate picture of the real gains or losses of a stock market's performance. Typically, returns are significantly reduced when one takes into account inflation and its negative impact on the purchasing power of money.

During the past decade, inflation in the advanced G-20 economies has averaged 2% a year and has ranged from a high of 3.2% in Australia to a low of -0.2% in Japan, an economy which has continued to struggle with deflation. Inflation was much more rampant in the emerging economies. As a group, prices in emerging markets increased on average by 8.5% a year in the 2000s. Moreover, unlike the advanced economies, the dispersion in inflation rates was much wider ranging from a low of 1.9% in China to a high of 23.1% in Turkey (see Table 1, column 3).

In the advanced economies, after adjusting for inflation, South Korea's market gained 21.9% and was followed by Australia and

Canada which both posted a ho-hum total gain of 13.8% over the ten-year period. The stock market losses delivered by the other advanced economies were also considerably magnified when one takes inflation into account. Total cumulative losses ranged from *minus* 57.5% in Italy's market to -42% in the US's S&P 500 to -27% in Germany's DAX index (see Table 1, columns 4 & 5).

In contrast, despite having to significantly counter higher inflation rates, nine of the ten stock exchanges of the emerging markets outpaced inflation and delivered positive real returns. After adjusting for inflation, total gains over the decade ranged from 74.1% for Indonesia's market to a high of 183.9% for Mexico's market. The only exception was the Turkish stock exchange where inflation wrecked its performance and turned a 248.6% nominal gain into a real loss of -64.4%.

The Bottom Line

No one could have predicted ten years ago that the stock markets of virtually all the advanced economies would inflict such horrendous losses on investors. Nor could one have foretold that investors in the emerging markets would be rewarded with such stellar gains.

Similarly, trying to predict what this decade will hold for stock market investors is nigh impossible and is best left to the legion of crystal ball gazers. All one can say with confidence is that, in the second decade of the 21st century, the world's stock markets will post returns that will be full of surprises both good and bad.

14

US Economy Faces A Precarious Recovery Path

(February 2010)

Headline GDP Number Masks Half-Hearted Recovery

The January 29 release of the fourth quarter 2009 GDP preliminary numbers by the Commerce Department confirmed that the US economy has emphatically pulled out of recession and a recovery is now well underway. After shrinking for four consecutive quarters, with output declining by 3.8% from its peak in Q2 2008 to its trough in Q2 2009, growth resumed in the second half of last year. What's more, after expanding at an annualized rate of 2.2% in Q3 2009, the pace of growth accelerated in the final three months of last year with GDP expanding by 5.7%.

While the headline GDP growth number has given the bulls a momentary victory in their claim that a V-shaped recovery was finally here, it is apparent that on a closer examination of the breakdown in GDP figures that this is a half-hearted recovery. Nearly three-fifths of the rise in GDP was accounted for by firms replenishing their inventories. But a look at the final sales to domestic purchasers numbers (which strips out the change in inventories and measures all the goods and services US residents have bought irrespective of where they were produced and gives a more accurate picture of the underlying strength of US demand) shows that final sales actually slipped to 1.7% in the fourth quarter, down from 2.3% in the previous three months.

In addition, it's interesting to note that in the first six months of the recovery following the deep recessions of 1973-75 and 1981-82, where GDP declined by 3.2% and 2.6% respectively, final sales to domestic purchasers surged by 4.7% and 6.5% compared to 2% in the current recovery (see Table 1).

TABLE 1
ACTUAL GAINS IN GDP AND FINAL SALES IN FIRST 6 MONTHS FOLLOWING MAJOR U.S. RECESSIONS

	Recession Trough Quarter		
	Q2 2009	Q4 1982	Q1 1975
GDP	4.0	7.2	5.0
Final Sales*	2.0	6.5	4.7

* Final Sales to Domestic Purchasers Equals (GDP + Imports – Exports – Change in Inventories).

Sources: NBER; Department of Commerce

Market Reaction – From Euphoria to Despondency

The initial reaction of the major US equity markets to the GDP numbers was positive but on sober reflection as investors started to digest the data and question the recovery's sustainability, the markets sold off. At the end of the day all ended in negative territory with the Dow Jones losing 0.5%, the S&P 500 1% and the NASDAQ shedding 1.5%.

The underlying worry among investors is that the recovery so far is being artificially driven by both the government's various stimulus measures including the 'cash for clunkers' program, where Americans trade in their old cars for new models, and tax credits for first time homebuyers, and the rebuilding of inventories. The assumption amongst US policy makers is that as these programs start to wind down and the boost from inventory rebuilding fades, consumer spending and business investment will pick up the slack and drive growth. But, that assumption could face some strong headwinds.

Factors Constraining US Growth to Persist

Faced with a myriad of problems, it is unlikely that the private sector will be healthy enough to be an engine for self-sustaining growth in the foreseeable future. Despite the resumption in growth, the economy continues to shed jobs. It lost a further 20,000 jobs in January and since the downturn began in December 2007, 8.4 million jobs have been eliminated. With the economy still shedding jobs, it goes without saying that Americans will not be able to ramp-up consumer spending.

Moreover, weighed down with a heavy debt-load and worried about the future, consumers are paying down their debts at a record pace. Latest data for December shows that consumer borrowing has dropped for an unprecedented eleven consecutive months. Consumers are not only borrowing less money, but they are also saving more. As a percent of disposable income, the savings rate has gone up from 1.7% in 2007 to 4.6% in 2009. In monetary terms, consumers saved $180 billion in 2007 but socked away $500 billion last year. Whether by choice or not, it is apparent that being thrifty is now becoming the 'new normal' for the American consumer.

So, until there is a sustained and meaningful contribution from consumption, which accounts for 70% of US GDP, this recovery is certain to be disappointing.

On the monetary policy front, to jump-start the economy the Federal Reserve has kept its benchmark interest rate near zero for the past fourteen months. But, this has failed to ignite demand. Credit conditions remain tight and the demand for both business and household loans continues to weaken. According to the Fed since December 2008, commercial and industrial loans have dropped to $1.32 trillion from $1.62 trillion, commercial real- estate loans have declined to $1.63 trillion from $1.73 trillion, and consumer loans have fallen to $814 billion from $861 billion. Without a pick-up in credit demand, the economy cannot expand and growth will stagnate.

Risk of a Double Dip Recession Remains

The above developments hardly point to an economy that is gearing up for a period of rapid growth. On the contrary, should these trends persist, there is a very real danger that the US economy could stumble again. After a quarter or two of positive growth, GDP could slide back into negative territory. Hence, one cannot categorically rule out the possibility of the US experiencing a double-dip recession.

Predicting the economic future is always hazardous but what we can be sure about is that the future path of growth will not be linear but will zigzag as the US economy struggles to get back to its long-term trend rate of growth.

A Patchy US Recovery Spells Trouble for the World Economy

An on-again/off-again US recovery does not bode well for the world economy. Despite China's high-speed growth over the past two decades and its rapid climb up the worlds GDP rankings, the United States remains by far the most dominant economy in the world. Based on the latest World Bank estimates the US accounts for about 24% of global GDP at current market exchange rates, and China 7%. Given its size and close trade linkages with the rest of the world, it is clear that a weak and halting US recovery will have a dampening impact on the growth rates of other countries.

In particular, the impact will be felt most keenly in those countries where the US is the main destination for their exports and also for those where the US accounts for a significant portion of their overall exports. Interestingly, although the US is the top destination for China's exports, it only accounts for less than 18% of its total exports.

Canada & Mexico Stand to Lose the Most

However, both Canada and Mexico, with their heavy reliance on the US market which accounts for about 80% of their total exports, stand to lose the most from a sub-par US recovery. While the North American Free Trade Agreement (NAFTA) cemented their access to the huge US domestic market and significantly boosted each country's exports in the past decade, the adjustment to lower demand by a crippled US economy will dent future growth in both Canada and Mexico.

The longer it takes for the US economy to fully recover, the harder it will be for Canada and Mexico to return to their full potential.

15

Greece's Agony

— An Economy on the Brink of Disaster —

(March 2010)

Greece's economy is rapidly spiraling out of control. Saddled with a budget deficit of close to 13% of GDP – the largest in the European Union – a debt-to-GDP ratio of over 110% and a current account deficit of around 10%, the chickens are now coming home to roost with a vengeance. After a decade of profligate spending, the country now faces a prolonged period of economic austerity, rising unemployment and falling living standards.

What precipitated the current crisis was the revelation last October by the new socialist government that the budget deficit was virtually double the previous estimate. Evidently, the preceding conservative government had been fudging the numbers. Predictably, market reaction to the news was swift. Investors started dumping Greek bonds on the market which sent yields surging, significantly raising the borrowing costs for the government. For example, prior to the crisis, the average yield on the 10-year Greek government bond was around 4.5%, compared to today's yield of about 6.25%.

In the immediate future, the government needs to borrow 20 billion Euros to refinance its bond holdings that are maturing in April and May. With investors continuing to be concerned that the government might default on its huge debt and coupled with the downgrades by the major rating agencies, it is a given that it will be considerably more expensive for the Government to borrow the funds it needs to keep going.

Government Hamstrung on the Economy Policy Front

To extricate the economy from its deep hole is going to be both challenging and extremely difficult. The problem is that in dealing with the economic crisis, the Greek government's hands are effectively tied. Being part of the European Union and sharing the common currency of the Euro, it cannot pursue an independent monetary policy; it cannot devalue the currency nor can it simply opt out of the Euro. At a minimum, pulling out of the euro would trigger another Lehman Brothers fiasco and send world financial markets reeling. It would also be 'lights out' for the fledgling global economic recovery.

On the fiscal policy front, with the Greek economy still firmly stuck in recession – GDP contracted for a fifth consecutive quarter in the fourth quarter of 2009 – the government cannot use its fiscal levers to pump-prime the economy. On the contrary, instead of being able to use fiscal policy to stimulate aggregate demand, the government is being forced to implement draconian fiscal measures to slash the deficit and reduce its national debt of close to 300 billion Euros. To this end, public sector salaries have been cut, various taxes are being increased, pensions have been frozen and the retirement age is being pushed-up from 61 to 63.

Austerity Measures to further Depress the Economy

The trouble is that the impact of all these austerity measures, while necessary and unavoidable, will not only further depress the economy but it will also prolong and deepen the recession. Nevertheless, despite these deflationary measures, the government is banking on the private sector to pull the economy out of its slump and return it to positive growth. That, however, will be a tall order. Sideswiped by the global recession, both consumer spending and business investment have been contracting for several months now and the latest spending cuts and tax hikes will only exacerbate the situation. Moreover, with borrowing costs rising, loan demand is likely to be moribund.

In addition, given the size of its current account deficit, the Greek economy is not competitive and thus it cannot rely on the export sector to bail it out. Not having the option to restore its competitiveness through currency devaluation, some analysts have suggested that the only way for Greece to boost its competitiveness is for the economy to lower its costs relative to its competitors. While in theory this makes sense, in practice it is a non-starter as it would require across the board wage cuts and a more flexible labour market and is something that the unions would fiercely resist.

But, despite the powerful headwinds facing the economy, the government nevertheless heroically expects GDP to contract by a negligible 0.3% this year, before returning to positive growth next year. Moreover, the forecast is for growth to accelerate in the ensuing three years rising from 1.5% in 2011 to 1.9% in 2012 to 2.5% in 2013. Given the severity of the underlying cyclical and structural issues that the country faces, these projections are overly optimistic and lack credibility.

Deficit Reduction Targets Unlikely to be Met

With the Greek economy still stuck in recession, it is highly unlikely that the government will be able to hit its deficit-reduction targets of 8.7% of GDP this year, let alone meet the EU benchmark of 3% in 2012. The austerity measures announced so far have only temporarily eased the financial crisis, but there is still a long way to go before the economy stabilizes.

Indeed, as noted above, the recession is likely to deepen under the weight of painful cuts and this will inevitably push back hopes of a much needed rebound. The problem is that, should the economy's performance worsen over the next few quarters, the deficit could well start to climb again. While the odds of such an outcome occurring may appear to be small, it would be foolish to dismiss it out of hand.

Given all the uncertainty about the near-term direction of the economy, investors and speculators betting against Athens achieving its budget targets will likely remain on the forefront for the foreseeable future. It may well take a lot more fiscal tightening to appease the bond market, but that would put the Greek government in an untenable situation forcing it either to seek a bail-out or default on its debt.

The Bottom Line

As Greeks stare into the economic abyss, it is little wonder that at a recent press conference the country's Prime Minister, Mr. Papandreou, said that '...the financial crisis threatens the (very) sovereignty of the country'. It is a melodramatic statement for sure, and one that was squarely aimed at preparing his fellow countrymen of the painful times that lie ahead. But, let us hope that it doesn't turn out to be prescient as well.

16

A Stronger Yuan No Cure For An Ailing US Economy

(April 2010)

Over the past several weeks a political war of words has been raging between the United States and China. At issue is the value of China's currency, the Yuan. Washington is accusing Beijing of effectively subsidizing its export industry by deliberately undervaluing its currency.

In an attempt to shield itself from the global financial crisis, China has pegged its currency at 6.83 to the dollar since July 2008, which Washington says gives its economy an unfair trade advantage. The contention is that this has contributed to US job losses and is the root cause behind America's huge trade deficit.

Faced with high unemployment and a fragile recovery, scores of US legislators are therefore pressuring the US Treasury Department to brand China as a 'currency manipulator' and, should China resist revaluing the Yuan, for the Obama administration to take action and impose trade restrictions including higher tariffs on Chinese-made imports.

Not surprisingly, China's President Hu Jintao has rejected these demands and has asserted that his country will not yield to outside pressure. The Chinese government's position is that it alone will proceed with currency reforms based on its own economic and social-development needs.

Thus, with the two big nations at loggerheads, the saber-rattling continues and, given all the rhetoric, the key question for investors is: will an appreciation of the Yuan actually rectify the US's trade imbalance with China?

China/US Trade – A Few Facts

To answer this question, a good starting point is to analyze the data on merchandise trade flows between China and the US. Table 1 highlights the trends in exports and imports between the two countries over the past five years and from this data a number of key points leap out.

TABLE 1 US/CHINA TRADE IN GOODS ($ billions)				
Year	US Exports to China	US Imports from China	Trade Balance	Import/Export Ratio
2009	69.6	296.4	-226.8	4.3
2008	69.7	337.8	-268.0	4.8
2007	62.9	321.4	-258.5	5.1
2006	53.7	287.8	-234.1	5.4
2005	41.2	243.5	-202.3	5.8

Source: US Census Bureau

First, US imports from China fell by 12.3% to $296.4 billion in 2009 as a result of declining US demand caused by the deep recession. However, despite the deceleration in China's growth rate from 13.0%

in 2007 to 8.7% last year, US exports to China have held their ground.

Second, US exports, although starting from a much lower base, have been rising at a faster pace than imports. This can be seen by examining the ratio between imports and exports (column 5). In 2005 for every dollar of goods that the US exported to China, it imported 5.8 dollars in return. However, by 2009 this ratio had fallen to 4.3 to 1 as the pace of exports accelerated.

Third, despite faster growth in exports, America's trade deficit with China has continued to rise. What's interesting to note is that for three of the five years during this period, the Yuan strengthened and the dollar depreciated. One would have thought that the realignment of the dollar/Yuan exchange rate would have had the intended effect of reducing the trade deficit. What gives?

The 2005 – 08 Revaluation of the Yuan

When China allowed the Yuan to appreciate by 22% against the dollar from 2005 and 2008, the resulting lower prices of US goods did indeed boost US exports to China which surged by 69%. In contrast, with the dollar price of Chinese goods becoming more expensive, although US imports during this period also increased it was by 'only' 39%.

In dollar terms, over this three year period, US exports to China increased by $28.5 billion, rising from $41.2 billion in 2005 to $69.7 billion in 2008. US imports from China on the other hand increased by $94.3 billion, rising from $243.5 billion to $337.8 billion, and swamped the increase in US exports. Thus, although the 22% hike in the value of the Yuan did alter the growth rates of exports and imports, because of the much lower base of US exports, the US trade deficit with China actually increased. In 2005, it was $202.3 billion but by 2008 it had jumped by nearly 33% to $268 billion.

A Stronger Yuan Will Not Bridge the Trade Gap

Certainly, a revaluation of the Yuan, by lowering US export prices and raising import prices, will alter the two-way trade flows between China and the United States. But, because of the wide gap that exists between what the US exports to China and what it imports, a revaluation of the Yuan will do little to correct the trade imbalance. In absolute terms China will still export more to the US and the trade deficit will continue to rise.

To put this in perspective, consider that from 2000 to 2009, US exports to China climbed by 18.2% a year while its imports increased by 13.7% annually during the last decade. Now, even if we assume that the US will be able to maintain these growth rates in exports and imports for the next 10 years, its trade deficit with China would still increase. In fact, it would jump from $202.3 billion in 2009 to $700 billion by 2019. Moreover, even if US imports from China were to remain frozen at the 2009 level of $296.4 billion and US exports continued rising at 18.2% a year, it would take nine years to achieve a balance in trade between the two countries.

Clearly, it is going to take a lot more than a currency revaluation of the Yuan to rectify the trade imbalance between China and the United States. Of course, a sharp appreciation of the Yuan in the order of 50% or more would quickly change the relative prices of imports and exports and reduce China's trade surplus with the US but this course of action is simply a non-starter.

China's concern is that if it were to allow its currency to float freely on foreign exchange markets, a sharp and sudden appreciation of the Yuan would rock its domestic economy and also de-stabilize the world economy. In China, millions of small-to-mid sized export-based companies that are already struggling with razor-thin profit margins would go under and send the unemployment rate soaring.

Some estimates suggest that about 30 million Chinese workers could be thrown out of work.

Furthermore, and contrary to Washington's claims, a sharply lower dollar would not lead to a resurgence in the US's manufacturing sector or higher employment. Instead, it is more than likely that the scores of US corporations that have manufacturing facilities located in China will simply shift their plants to more lower cost locations such as India, Indonesia, or Vietnam.

In addition, higher import prices would further squeeze the incomes of millions of US consumers who are struggling to stay afloat. Inevitably, this will lead to a slowdown in consumer spending and effectively put the nascent US recovery in jeopardy.

The Bottom Line

Of course, China recognizes that it is in its self-interest to let the Yuan rise in value. A stronger currency would increase the purchasing power of its citizens; falling import prices would help curtail rising inflation and give the central bank more leeway in conducting its monetary policy; and, by increasing domestic demand, a stronger Yuan would help China rebalance its economy. But, China will manage the value of its currency on its own terms and it will do so in a deliberate and measured manner.

Letting the Yuan appreciate against the dollar will certainly help ease the pressure between the two countries, but, as noted above, it will not solve the trade imbalance. For that to happen, the US needs to reduce its level of consumption and step up its savings rate. But, with consumption accounting for 70% of GDP, reducing consumer spending risks pushing the US economy into a prolonged period of stagnation.

Ironically, the adjustment to a stronger Yuan will be much more wrenching for the United States than it will be for China. Perhaps Washington needs to reflect on the old adage: *Be careful what you wish for*....

17

Soaring G7 Government Debt And The Interest Rate Paradox

(May 2010)

While the focus of investors is currently on the debt crises that are plaguing Greece, Spain and Portugal, it is important not to forget – lest one gets blindsided – that all the major economies of the G7 also face serious budgetary and debt problems. To date, G7 governments have benefited enormously from the prevailing low interest rates and, despite soaring deficits and debt levels, the cost of servicing the debt has actually been declining.

But, as investor angst about the fragile state of the global recovery spreads and deepens, this sweet spot could come to an abrupt end. Jittery credit markets are likely to push up the cost of credit and this will make life much more difficult for governments as they try to grapple with their burgeoning debt levels.

Budget Deficits Surge

Before the financial crisis hit the global economy with full force in 2008, the fiscal situation in the G7 raised few eyebrows. In 2007, the overall fiscal deficit of the G7 amounted to 2.1% of GDP. At the individual country level, both Canada and Germany actually had budget surpluses and although Britain, France, and the US had the largest budget shortfalls, at 2.7% of GDP, they were relatively non-toxic. But, the severe global economic downturn coupled with government stimulus measures aimed at shoring up their respective economies has wrecked havoc on the fiscal situation (see Table 1).

Within the space of only two years, the overall deficit in the G7 has quintupled and reached 10% of GDP last year. All G7 members are now running deficits and the deterioration is indeed eye-popping. The US's budget deficit jumped from 2.7% of GDP in 2007 to an alarming 12.5% last year and the fiscal situation has also plummeted sharply in Britain and Japan with both economies recording double-digit deficits in 2009.

TABLE 1
BUDGET DEFICITS
(% of GDP)

G-7	2007	2009	2010
United States	-2.7	-12.5	-11.0
Canada	+1.6	-5.1	-5.2
Japan	-2.4	-10.3	-9.8
Britain	-2.7	-10.9	-11.4
France	-2.7	-7.9	-8.2
Germany	+0.2	-3.3	-5.7
Italy	-1.5	-5.3	-5.2
Average	-2.1	-10.0	-9.5

Source: IMF Fiscal Monitor, May 14, 2010

Moreover, with the economic recovery still shaky in many G7 economies, the overall fiscal balances are expected to improve only marginally in 2010. In its latest projections, the IMF expects the deficit to trend higher this year in Britain, France, Germany, and Canada but to dip in the US, Japan, and Italy.

Getting Buried in Debt

Saddled with substantial budget deficits, the accumulated national debt levels have also been rising dramatically in all the G7 countries. The overall debt-to-GDP ratio has jumped by 20 percentage points rising from a pre-crisis level of 82.2% in 2007 to 102.3% in 2009. Moreover, this figure is projected to climb further to 110.2% this year (see Table 2). For sure, a number of countries began the crisis with high debt ratios and here Japan is the stand-out. With a debt-to-GDP ratio of 217.7% in 2009, its national debt is now larger than the combined annual output of Britain, France, Germany and Italy.

TABLE 2 GROSS GOVERNMENT DEBT (% of GDP)				
G-7	2007	2009	2010	CAGR[1/]
United States	62.1	83.2	92.6	14.3%
Canada	65.0	82.5	83.3	8.6%
Japan	187.7	217.7	227.1	6.6%
Britain	44.1	68.2	78.2	21.0%
France	63.8	77.4	84.2	9.7%
Germany	65.0	72.5	76.7	5.7%
Italy	103.4	115.8	118.6	4.7%
Average	82.2	102.3	110.2	10.3%

[1/] Compound annual growth rate from 2007-2010
Source: IMF Fiscal Monitor, May 14, 2010

While debt levels have shot up in all the major economies, the rate at which the debt-to-GDP ratio is rising varies considerably. As you can see from Table 2, column 4, the overall national debt in the G7 is growing by 10.3% a year. But what is indeed startling is that Britain's national debt is increasing by 21% a year and that of the United States by 14.3%.

Bond Yields Falling in G7 Economies

Incredulously, despite all the hand-wringing by credit markets regarding the sustainability of zooming deficits and exploding debt-levels in the G7 nations, interest rates on government bonds actually have been falling. Since the onset of the financial crisis, yields on 10-year government bonds are now lower in *all* the major advanced economies (see Table 3). Moreover, rates have continued to fall despite the debt crisis that has gripped the Eurozone economies of Greece, Spain, and Portugal. In these economies, investors, reacting to ratings downgrades and concerned about the risk of default, have sold off government bonds. This in turn has sent bond prices tumbling and pushed up yields. (Bond prices and yields move in opposite directions).

TABLE 3
10-YEAR GOVERNMENT BOND YIELDS
(Percent)

G-7	Dec 31 2007	Dec 31 2009	May 25 2010	YTD % point Chg
United States	4.0	3.8	3.1	-0.7%
Canada	4.0	3.6	3.2	-0.4%
Japan	1.5	1.3	1.2	-0.1%
Britain	4.6	4.0	3.5	-0.5%
France	4.4	3.6	2.9	-0.7%
Germany	4.3	3.4	2.6	-0.8%
Italy	4.6	4.2	4.0	-0.2%

Source: Thomson Reuters

This of course raises a key question – why would government bond yields be falling across the board in the G7 when national debt levels are spiraling out of control? It certainly seems paradoxical but there are a couple of plausible reasons for this.

Firstly, and most importantly, it is primarily the result of quantitative easing and the aggressive purchase of government bonds undertaken by the major central banks which have helped to hold down rates.

Secondly, the recent sharp falls in world stock markets have triggered a flight to safety as investors have flocked to the security of government bonds. As a result, the bond market has rallied sending yields down.

Get Ready for the Coming Bond Bear Market

However, this state of affairs cannot last forever. Indeed, the gigantic financing requirements needed by the G7 governments to fund their budget deficits means that a huge wave of bonds will be continuously coming on to the market place. To put this in perspective, before the financial crisis, the overall deficit amounted to about $630 billion. But by last year, this figure had ballooned to over $3 trillion.

As the global economy emerges from its deepest slump since the Great Depression of the 1930s, central banks have signaled that they intend to start the process of 'normalizing' monetary policy by withdrawing the 'emergency' stimulus measures. As they unwind their quantitative easing policies and scale back on their bond-purchase programs the onus will be on the private sector to digest the issuance of government bonds to fund the deficit.

But here's the rub. With insufficient private sector savings to absorb the ever-growing supply of bonds this means that interest rates have only one way to go, and that is up.

One thing is for sure; it is going to take several years of fiscal austerity to bridge the yawning gap between government revenues and spending. As governments start to tighten fiscal policy and lay out their multi-year plans to rein in their budget deficits, to compensate, central banks are likely to keep monetary policy quite lax. While this is the right policy mix to engender economic growth, it may unfortunately prove to be ineffective. The reason; the need to finance persistently high national debt levels by all the G7 economies will lead to rising bond yields.

The Bottom Line

We may be entering a period where, notwithstanding the intentions of central bankers, monetary policy effectively tightens. The danger here is that a combination of tight fiscal and monetary policy could effectively flatten the recovery.

Under these circumstances, investors would be wise to take note that government bonds are no longer the safe haven asset class they once were and it is probably expedient to begin preparing for the coming bear market in bonds which is just around the corner.

18

From Recession To Recovery In The G7

— *An Early Assessment* —

(June 2010)

Thankfully, the global recession that rocked the world economy in 2008 and 2009 and dealt a savage blow to several advanced economies is now history. Although economic growth has resumed in all the major economies of the world the road ahead will not be an easy one. In particular, investors continue to fret that unless governments take control of their burgeoning budget deficits and soaring public sector debt levels, a second down leg of the Great Recession is inevitable.

However, as the focus of government policy makers is now on economic recovery, this seems an appropriate time to evaluate the damage the recession inflicted on the major advanced economies and gauge the strength of the recovery in each of the G7 countries.

Recession's Impact on GDP

Table 1 highlights the impact of the recession on GDP for the major economies. As you can see, the impact of the slump on the overall level of output varied considerably across the G7 hitting some countries much harder than others. The length of the recession also varied but the United Kingdom was the stand-out; it experienced the longest recession which lasted for 18 months.

TABLE 1
DEPTH & DURATION OF THE RECESSION IN G7 ECONOMIES

	GDP Peak Quarter	GDP Trough Quarter	Quarters in Recession[1]	Cumulative Decline in GDP[2]
Canada*	Q4 07	Q2 09	5	-3.4%
France	Q1 08	Q1 09	4	-3.9%
Germany	Q1 08	Q1 09	4	-6.7%
Italy	Q1 08	Q2 09	5	-6.7%
Japan	Q1 08	Q1 09	4	-8.6%
Britain	Q1 08	Q3 09	6	-6.3%
United States	Q2 08	Q2 09	4	-3.8%

*Canada's economy recorded small back-to-back falls in GDP in the first two quarters of 2008, rebounded somewhat in the third quarter, and then contracted for three consecutive quarters before finally hitting bottom in the second quarter of 2009.

[1] Number of negative quarters; [2] From peak to trough
Source: OECD

Canada's recession was less severe than other G7 countries. Between the fourth quarter of 2007 and the third quarter of 2009, real GDP fell by a cumulative 3.4%. In contrast, Japan, the world's second largest economy, contracted by 8.6% over a period of four quarters. Similarly, Europe's big economies were also hit pretty hard with Germany, Italy and the United Kingdom suffering output declines of more than 6%.

But what's particularly remarkable to note is that the United States, despite being at the epicenter of the financial crisis which triggered the global downturn, experienced an overall decline in output of 'only' 3.8%. With the exception of Canada and France, this was considerably less than the declines recorded by the other G7 countries.

An Uneven Growth Path but Canada Alone has a V-Shaped Recovery

When it comes to the recovery, the economies of Germany, France and Japan resumed growing in the second quarter of 2009. Canada and the US pulled out of recession in the third quarter while the UK economy returned to growth in the final three months of 2009 and was the last G7 member to exit recession (see Table 2).

TABLE 2
THE UNEVEN ROAD TO RECOVERY

	Start of Recovery	Quarters in Recovery	GDP Growth Cumulative	GDP Growth Annualized	% from previous peak*
Canada	Q3 09	3	2.9%	3.9%	-0.5%
France	Q2 09	4	1.2%	1.2%	-2.8%
Germany	Q2 09	4	1.5%	1.5%	-5.3%
Italy	Q3 09	3	0.7%	1.0%	-6.1%
Japan	Q2 09	4	4.2%	4.2%	-4.7%
Britain	Q4 09	2	0.7%	0.9%	-5.5%
USA	Q3 09	3	2.7%	3.6%	-1.2%

* From start of recovery to 2010 Q1; Source: OECD

As in the case of the recession, the pace of the recovery has also varied considerably. On a comparable basis, the annualized rate of growth in GDP has ranged from a low of 0.7% in the UK to a high of 4.2% in Japan.

Interestingly, since pulling out of recession the pace of economic growth in Canada has accelerated and it is the only major economy that is experiencing a 'V' shaped recovery. The country's GDP growth has speeded up, going from 0.9% in Q3 2009, to 4.9% in Q4 2009 to 6.1% in Q1 2010.

In contrast, the latest GDP growth numbers show that in several G7 countries there was a marked slowdown in the recovery in the first quarter of 2010. For example, growth in the United States decelerated from 5.6% in Q4 2009 to 3.0% in Q1 2010. During the same period, growth in France fell from 2.2% to 0.5% and in the United Kingdom from 1.8% to 1.2%.

Returning to Pre-Recession Levels – A Likely Timeline

Despite the return to growth, the level of output in several G7 economies nevertheless remains significantly below pre-crisis levels. For example, output in Italy is still 6.1% below its pre-recession level. Similarly, the economies of the United Kingdom and Germany are 5.5% and 5.3% smaller than they were at the start of the recession two years ago. And although Japan's GDP is up 4.2% from the bottom, it is still off 4.7% from the pre-recession peak (see Table 2, column 6).

In sharp contrast to Europe and Japan, output in North America is now nearly back at pre-recession levels. Canada's overall GDP is currently only 0.5% below its pre-recession peak and the comparable figure for the US is 1.2%. While growth is projected to moderate in the second half of 2010, both North America economies will most likely transition from recovery to expansion later this year.

But, as noted above, the same cannot be said for Europe and Japan. It will take them longer to get back to pre-recession levels of economic activity. Indeed, based on the latest (April 2010) growth projections of the International Monetary Fund, France, Germany, the UK and Japan are not expected to return to pre-recession levels of real GDP until sometime in late 2011 or 2012 at the earliest. And that's assuming that the forecasts are bang on target

The Bottom Line

Although the G7 economies are continuing to recover from recession, the upturn remains tentative and uneven. In particular, the economic recovery in Europe is at risk of coming unglued. The sovereign debt crisis that is currently plaguing Greece, Portugal and Spain remains a major threat and one that could yet engulf other European economies and stall global growth.

To date, the recovery has been supported by the unprecedented levels of government spending which, while preventing a repeat of the Great Depression, has sent budget deficits and government debt levels soaring. But, as the support from fiscal stimulus starts to fade and policy makers begin to tighten fiscal policy, the public sector will no longer be the main driver of the economy.

The onus will now increasingly be on the private sector in the G7 countries to carry the baton. However, unless the other main components of GDP, including business investment and exports, vigorously kick in, a self-sustaining recovery will remain elusive. Unfortunately, the pain that the recession inflicted on the major advanced economies is not expected to end anytime soon.

19

The Bears Are Circling Again

(July 2010)

Contrary to expectations, global equity markets are getting battered again as the roller coaster ride in stocks continues. After crashing in 2008 as the slump in the world economy deepened, stock markets, in anticipation of a global recovery in both developed and emerging economies, reversed course in 2009 and posted double-digit gains. And now that a global recovery is underway, markets have reversed course yet again and are heading back into bear market territory. It's enough to give even the most seasoned investor a full-blown case of the jitters.

Stock Price Declines are Universal

As one can see from the performance of the various stock exchanges listed in Table 1, the decline is universal. Following the commonly accepted definition of a market correction (a drop of at least 10%) and a bear market (a drop of 20% or more), a quick headcount shows that 12 of the 19 markets are already in correction territory and three (China, Greece and Spain) are deep into a bear market.

Among the advanced G-20 economies, 8 of the 11 major exchanges posted double-digit losses from their respective 2010 highs. South Korea's Kospi Index, with a loss of 3.1%, was the best performer while exchanges in Italy and Japan brought up the rear with each posting a decline of 18.9%.

	Index	Market Peak in 2010	% Decline from Peak
TABLE 1 — GLOBAL STOCK MARKETS HEADING INTO BEAR MARKET TERRITORY			
Advanced G-20 Economies			
Australia	All Ordinaries	15 Apr	-13.9%
Canada	S&P/TSX	26 Apr	-8.0%
France	CAC 40	15 Apr	-15.3%
Germany	DAX	26 Apr	-5.8%
Italy	FTSE/MIB	8 Jan	-18.9%
Japan	Nikkei 225	5 Apr	-18.9%
South Korea	Kospi	26 Apr	-3.1%
Britain	FTSE 100	15 Apr	-15.6%
United States	DJIA	26 Apr	-12.8%
United States	S&P 500	23 Apr	-15.3%
United States	Nasdaq	23 Apr	-16.6%
BRIC Economies			
Brazil	Bovespa	8 Apr	-15.1%
Russia	Micex	15 Apr	-14.5%
India	Sensex 30	7 Apr	-1.5%
China	Shanghai Com	5 Jan	-26.9%
PIGS Economies			
Portugal	PSI General	8 Jan	-17.1%
Ireland	Irish Overall	26 Apr	-17.7%
Greece	FTSE/ASE 20	8 Jan	-43.9%
Spain	IBEX 35	6 Jan	-24.2%

Source: Bloomberg

All the stock markets of the so-called PIGS economies have posted double-digit losses. The largest decline was recorded by Greece

(-43.9%) followed by Spain (-24.2%), Ireland (-17.7%) and Portugal (-17.1%).

As for the BRIC economies, the Bombay Stock Exchange's Sensex Index was the best performer, losing 'only' 1.5% from its high for the year in April. At the other extreme, China's stock exchange has **dropped by 26.9% from its January high. Indeed, the drop in the Shanghai Composite Index, when measured from its record high of** 6092.06 reached on October 16, 2007 to its June 30, 2010 value of 2398.37 is truly an attention-grabber; the index is down a gut-wrenching 60.6%.

What's Behind the Markets Latest Drop?

There are several reasons why markets have been tumbling around the world. In particular, a string of downbeat economic reports in the past few weeks have shaken investor confidence about the sustainability of the global recovery.

Among the major economies the latest information shows that the US economy is weakening. Job growth in the private sector is fading, the pace of manufacturing activity is slowing, the housing sector remains in the doldrums, and with nearly one out of five mortgage borrowers underwater, US consumers remain in a precarious state. In addition, credit conditions are still restrictive. Indeed, according to the Federal Reserve, bank lending continues to contract – it fell in April and May at an annualized rate of 7.75%.

Economic growth is also slowing in China which has, up until now, been a powerful driver of global growth. Moreover, there is growing concern that the country's property market is in bubble territory and could start to deflate anytime now. Property prices fell 0.1% in June compared with May and it was the first monthly decline since February 2009.

In Europe, the sovereign debt crisis that is plaguing Greece, Spain, Portugal, and Italy has also alarmed investors. While there is general agreement that debt levels have to be reined in, the fear is that, with a recovery that is still very tenuous, the implementation of fiscal austerity measures in the big Eurozone economies of France and Germany as well as the United Kingdom is likely to squash growth in the region again and put the brakes on the global recovery.

G20 Toronto Summit Fails to Stabilize Markets

Any hopes that the June G20 Toronto meeting would help rescue the global economy also quickly evaporated. Although a deal to halve their respective deficits by 2013 and stabilize or reduce government debt-to-GDP ratios by 2016 was reached by the group, it was not binding and did nothing to stabilize financial markets.

Instead investors focused on the major policy split particularly between the US and the European Union concerning the speed with which to tackle budget deficits. The EU essentially favours big cuts now arguing that it will restore confidence in financial markets and spur growth. On the flip side, the opposing camp led by the US contends that economic stimulus measures are still needed in the short term to prevent the faltering recovery from stalling or even going into reverse.

With discord on the policy front, this has heightened investor fears that governments, instead of pointing the way forward, are completely baffled about how to solve the economic gloom that is again gripping the global economy. As investors gave the G20 communiqué a blanket thumbs down, markets resumed their decline.

More Stock Market Turmoil Lies Ahead

The global recovery is still a long way from being totally secure. Despite the sharp loosening of monetary policy by the major central banks, record low interest rates have not been able to lift demand in any meaningful way. On the contrary, business credit demand has been shrinking and there is growing evidence that the money supply is declining which usually is a sign of an impending slowdown in economic activity – certainly not a recovery.

On the fiscal policy side, the sharp ramping up of government spending did manage to prevent a complete collapse of the global economy. Although growth has resumed in all the major advanced economies, it remains extremely fragile and is still dependent on the government sector. The trouble is that the hoped for revival of the private sector as the major contributor to the recovery has yet to materialize.

The fear is that tightening fiscal policy now will knock global demand and could well push a number of economies back into recession – the dreaded double-dip – and send the unemployment rate climbing again. Moreover, under such a scenario, tax revenues are likely to fall which would in turn push up deficits. Certainly, countries do need to bring their deficits under control, but too much austerity too soon could choke off a fragile recovery.

The Bottom Line

As investor angst mounts and the bears tighten their grip on global equity markets, get ready for a market trajectory that is best characterized as one step up, two steps down. With the current policy discord and the heightened uncertainty about the sustainability of the recovery, it's a given that equity markets will continue to be volatile in the coming months.

But, all will not be downhill. Markets will also be punctuated by short-term technical rallies as investors jump on any 'good news' signs of recovery based on a single month's data flow. Clearly, however, this is not a market for the faint-of-heart.

20

The Curious Case Of Falling Bond Yields In G7 Economies

(September 2010)

One of the more intriguing developments in the global economy this year has been the divergent response of bond markets to swelling budget deficits and the sharp rise in government debt levels. Specifically, bond yields have continued to fall across the board in all the G7 countries but have increased in the peripheral economies of Portugal, Ireland, Greece, and Spain, the so-called PIGS.

Since the beginning of this year, 10-year government bond yields have fallen on average by 93 basis points (-27.4%) in the G7 nations and has ranged from a low of 31 bps in the case of Japan to a high of 136 bps in the United States.

In stark contrast, bond yields have increased on average by 204 bps (+38.6%) in the PIGS economies. The rise in rates has ranged from a low of 6bps in the case of Spain to a whopping high of 578 bps in Greece (see Table 1).

The question, of course, is why the divergence? After all, each country is saddled with rising budget deficits and soaring debt-to-

GDP levels. Indeed, according to the IMF, next to Ireland, the budget deficits in both Britain and the US are expected to top 11% this year. Moreover, compared to Japan's debt-to-GDP ratio of 227.1%, Greece's debt ratio of 133.2% of GDP looks positively paltry.

TABLE 1
TRACKING GOVERNMENT BOND YEILDS IN THE G-7 AND PIGS ECONOMIES

	31 Aug 2010	31 Dec 2009	Basis Point Change[1/]	YTD % Change
G-7 Economies				
United States	2.48	3.84	-136 bps	-35.4%
Japan	0.98	1.29	-31 bps	-24.0%
Germany	2.11	3.40	-129 bps	-37.9%
France	2.46	3.61	-115 bps	-31.9%
United Kingdom	2.83	4.01	-118 bps	-29.4%
Italy	3.77	4.16	-39 bps	-9.4%
Canada	2.76	3.61	-85 bps	-23.4%
PIGS Economies				
Portugal	5.48	4.06	+142 bps	+35.0%
Ireland	5.77	4.88	+89 bps	+18.2%
Greece	11.59	5.81	+578 bps	+99.5%
Spain	4.05	3.99	+6 bps	+1.5%

[1/] One basis point is equal to 1/100th of 1%, or 0.01%. Source: Thomson Reuters

Yet, interest rates in Japan have fallen by nearly a third whereas they have doubled in Greece (see Table 2). Of course, an obvious reason is that the risk of default in the PIGS is perceived, rightly or wrongly, by investors to be much higher than that of the G7 countries.

Is the Bond Market Ahead of the Curve?

However, given the divergence in yields, could the bond market actually be ahead of the curve and sending a signal that the biggest risk now confronting the G7 economies is deflation and not inflation? With 10-year bond yields at near-record low levels in all the G7 countries, the fear amongst investors is that rates could suddenly start to climb in the near future.

TABLE 2
BUDGET DEFICITS & GOVERNMENT DEBT IN THE G7 AND PIGS ECONOMIES
(In percent of GDP)

	Deficit	Debt
G-7 Economies		
United States	-11.0	92.6
Japan	-9.8	227.1
Germany	-5.7	76.7
France	-8.2	84.2
United Kingdom	-11.4	78.2
Italy	-5.2	118.6
Canada	-5.2	83.3
PIGS Economies		
Portugal	-8.8	86.6
Ireland	-12.2	78.8
Greece	-8.1	133.2
Spain	-10.4	66.9

Source: IMF Fiscal Monitor

The concern is that, as governments start to rein in spending while the recovery is still quite shaky, it will make the task of curbing the deficit much more difficult. The danger here is that over-zealous fiscal consolidation at this stage of the business cycle could snuff out the recovery and negatively impact the tax base.

Under such a scenario, with tax revenues declining, governments would then be forced to borrow even more to finance their deficits and meet their debt repayment obligations. At some point, the demand for government bonds will reach a saturation point and lead to higher interest rates.

Why Are Bond Yields Falling?

Amazingly, although the supply of government bonds in the G7 economies is skyrocketing, the demand for them has been even more voracious. As a consequence, with demand outstripping supply, the bond market has rallied strongly sending prices up and yields spiraling downwards.

A key reason why government bond yields are at record lows, and indeed may even go lower, is that investors are now much more concerned that deflation, with all its negative connotations, may be about to take hold. Until recently, the general view among analysts and investors was that interest rates would be trending up in 2010 as the recovery in the advanced economies gathered pace pushing-up inflationary expectations. But, this has not happened and the latest flow of economic data is anything but encouraging.

Economic Outlook Takes a Turn for the Worse

What is interesting is that despite massive government spending and the ultra-loose monetary policy pursued by the central banks over the past 18 to 24 months there is, as yet, little sign of a build-up of inflationary pressures. To be sure, the extraordinary policy response by governments did manage to prevent the Great Recession of 2008-9 from turning into a full-blown depression. But, it has failed to lead to self-sustaining growth and now, with the initial high of the policy response starting to wear off, growth is dropping off again.

Second quarter GDP figures show that growth in the US and Canada is slowing sharply and that Japan's economy has virtually stalled. In Europe, unemployment remains stuck at 10 percent and growth is also sputtering. As a consequence of these trends, the focus now among investors has switched from worrying about inflation to fretting about weakening growth and deflation.

Given that on the economic front there is zero evidence of any build-up of inflationary pressures in the G7, interest rates could remain low for quite some time. But, how realistic is this? In a world where budget deficits are rising sharply and debt-to-GDP ratios are zooming upwards, it is clear that interest rates cannot remain at these abnormally low levels indefinitely.

In reality, as nations compete for a limited pool of funds to finance their respective deficits, interest rates will likely lurch upwards. Indeed, we may be about to enter a unique period where economic growth essentially stagnates for a prolonged period but where interest rates, because of governments' huge borrowing needs, nevertheless start to climb.

The Impotence of Monetary Policy

The ultra-loose monetary policy that the G7 central banks have pursued in the past two years has, simply put, failed to lead to a resurgence in loan demand by corporations and households. On the contrary, the focus continues to be on paying down debt and, given that household debt-levels are still high, the deleveraging process is far from over.

In any case, with policy rates at rock bottom levels, the major central banks have now effectively run out of options as they have little room to lower interest rates further. Moreover, it would not do any good as record low interest rates over the past two years have failed to reignite demand. Extending the policy of quantitative easing whereby central banks buy up government and corporate debt directly will also have – at best – a marginal impact.

Fiscal Policy Tightening

As governments start to shift their focus from borrowing heavily to finance their spending to tax hikes and spending cuts in order to rein in ballooning budget deficits, means that tighter fiscal policy will further depress the economic outlook and could snuff out any flickering signs of economic revival.

The Bottom Line

From the evidence, one can't help but conclude that falling government bond yields are sending a strong signal that deflationary forces may be lurking around the corner. Of course, there's always the chance that the bond markets could have gotten it wrong. But, if not, the major advanced economies are headed for another fall.

With GDP growth already skidding in the G7 economies and with little room left on the policy front for governments to buttress their respective economies, the outlook is shrouded in uncertainty. Unfortunately, with growth projected to be erratic and unemployment expected to stay high, the economic future for the advanced countries looks quite painful.

21

From Recession To Depression

— The Wrenching Case of Ireland —

(October 2010)

Ireland's economy, once the star of Europe, is now in dire straits and the future unfortunately also looks quite bleak for the country. However, what is astonishing is the rapidity with which the country has gone from being one of the most prosperous economies in the European Union to one of its most desperate.

After averaging growth of over 7% a year from 1997 to 2007 – the fastest among all of the 30 plus OECD countries during this period – the economy has now gone into a tailspin and remains in a deep funk. In other words, it has virtually fallen apart.

This sorry state of affairs was brought on by a combination of events including excessive borrowing by households and some would say the reckless lending by Irish banks, the implosion of the property market, and the global recession. As one wag has poignantly observed, the Celtic Tiger has been declawed.

Ireland's Economy in Free-Fall

Table 1 highlights the devastating impact the recession has had on total output and the major components of the country's GDP. From the beginning of the downturn in the fourth quarter of 2007 to the second quarter of this year, output has contracted by gut-wrenching 13.5%.

The domestic economy has been especially hard hit. Since the onset of the recession, consumer spending has declined by 17.0% and investment spending has collapsed, falling by 48.7%. Moreover, on the back of budgetary cutbacks as a result of the deficit, government spending has also declined, further pulling down the economy.

TABLE 1 IRELAND'S ECONOMY IN FREE-FALL Q4 2007 – Q2 2010	
Personal Consumption Expenditures	-17.0%
Gross Investment	-48.7%
Government Expenditures	-9.8%
Exports	+1.2%
Imports	-11.1%
GDP	-13.5%

Source: Central Statistics Office Ireland

Fortunately, the export sector, which is dominated by foreign multinationals and is a key component of Ireland's economy, has so far managed to hold its own. After declining by over 10% during the worst period of the global recession in 2008, the export sector has managed to stage a come-back.

Boosted in part by the (earlier) decline in the Euro and the pick-up in global growth, exports have now increased for three consecutive quarters and are up by 12.1% from its low point reached in the third quarter of 2009. Although the export sector has regained its pre-recession peak level, the current slowdown in global growth is likely to stymie future growth. Indeed, the latest data release shows that Irish exports fell by 4% in August from a month earlier.

Aside from the latest information on the national accounts, the incoming flow of other economic data remains dismal. Unemployment in the country has jumped from 4.5% in 2007 to 13.6% in the second quarter of 2010. Out of a labour force of about 2 million, over 280,000 people have lost their jobs since the start of the recession in the fourth quarter of 2007. Moreover, employment levels have been falling for every single quarter since then.

Unsurprisingly, given the dire job market and being fearful of their jobs, Ireland's consumers, already laden down with high levels of household debt, have been battening down the hatches and shoring up their savings. As a result, the personal savings rate has shot up and, as a percent of disposable income, it has jumped from 1.7% in 2007 to 10% this year.

On the property front, according to the latest report published by the property website Daft.ie, the average asking price for houses has fallen by 37% since its 2007 peak and now stands at €195,000. Moreover, on a monthly basis, prices have now fallen for 33 consecutive months and there are little let-up in sight.

Government Policy Makers Boxed-In

The Irish Government finds itself entrapped on the economic policy front with little room to manoeuvre as it tries to extricate itself from its deep slump. As monetary policy is controlled by the European Central Bank (ECB), it can't lower interest rates nor can it engage in a round of quantitative easing to jump start its economy. On the exchange rate front, being a member of the Euro also rules out the option of devaluing the currency in order to boost exports.

Fiscal policy remains the only macro-economic policy weapon that the government has at its disposal. But even here, with the deplorable state of the government's finances, policy makers cannot cut taxes nor increase government spending. On the contrary, although the economy is experiencing a severe recession, the government in Dublin is nevertheless set to announce a further set of austerity measures later this year as it struggles to bring the country's massive budget deficit and mountainous public debt levels under control.

In just two years, Ireland has gone from having a balanced budget in 2007 to a deficit equivalent to 14.4% of its GDP last year. It now has the dubious distinction of having the highest budget deficit in the EU. To make matters worse, the government has just announced that the estimated cost of bailing out the country's ailing banks could now rise to over €34 billion (or over $47 billion) which is equivalent to more than a fifth of the country's national output. This of course has raised alarm bells in financial markets with investors questioning the country's ability to honour its debt.

Concerned about Ireland's deteriorating economy, all the major credit ratings agencies including Standard & Poors, Moodys and Fitch have already downgraded Irish debt and also have put the country on negative watch. This has pushed up bond yields, raising debt servicing costs and is a further constraint on the economy.

Ireland's Upcoming Budget to Knock the Wind out of its Economy

Ireland is caught between a rock and a hard place and, unfortunately, the situation is set to get even worse. With an eye to appeasing financial markets and meeting the EU's Stability and Growth Pact, the government has pledged to bring its deficit down from over 14% this year to under 3% by 2014. The upcoming budget is expected to contain between €5 and €10 billion in spending cuts and tax hikes that are to be spread out over the next four years.

The trouble is that, in trying to bring the deficit under control when the economy is contracting, this high-risk strategy could easily backfire. Instead of boosting the confidence of financial markets and pointing the way forward for Ireland's citizens, the recession is likely to deepen as domestic demand further retrenches. The very real danger here is that this could lead to a vicious circle whereby falling tax revenues pushes up the deficit which in turn engenders further austerity measures.

While the country clearly needs to bring its finances under control, a better course of action would be to extend the deficit-reduction timeline from four years to, say, six or eight years. Nevertheless, with economic policy set to tighten even further, it will take a miracle for the Irish economy to pull itself out of its deflationary slump.

The Bottom Line

The call for strong action and tougher measures to cut the deficit in order to appease the bond markets and ensure financial stability in Ireland is likely to flop again. Further austerity measures when the economy is already contracting will do nothing to put the country on the road to recovery.

Bond yields may fall temporarily but, should the economy lurch downwards, which is likely to happen, bond holders will be looking for the exit signs on concerns that a debt default or restructuring is inevitable. Given these developments, the odds are high that the Irish government will, sooner or later, be forced to seek the assistance of the IMF and the EU.

In terms of the bigger picture, the strait jacket that Ireland finds itself in can be seen as a harbinger of things to come for other heavily indebted EU economies including Spain, Portugal, Italy and Greece. Unable to formulate their own independent monetary and exchange rate policies to lift their respective economies from the doldrums, and finding themselves hamstrung by onerous debt levels, the spectre of recurring debt restructuring and/or outright debt defaults will haunt financial markets.

On the currency front, unless there is a dramatic improvement in the economic performance of the Eurozone as a whole, but especially in the smaller peripheral ones such as Ireland and Greece, the citizenry of these countries will increasingly question the wisdom of continuing with the Euro.

What was once simply an unthinkable proposition that was dismissed out of hand, is now finding its way onto the discussion tables as a possible solution. Make no mistake, the very viability of the Euro as the common currency for the 16 EU member countries is set to be severely tested in the coming months.

22

The Federal Reserve & QE2
— *A Hazardous Strategy Fraught With Pitfalls* —

(November 2010)

The US Federal Reserve, concerned about a flagging US economy and worried about the potential for deflation, announced on November 3 a second round of quantitative easing that has been dubbed QE2. The Fed's plan is to inject new cash into financial markets by buying $600 billion of Treasury bonds, which amounts to 4% of the nation's GDP, between now and the end of June 2011.

The Fed's decision to restart quantitative easing – effectively printing dollars to buy up US government debt – is intended to help boost economic growth, increase inflation and lower unemployment in the country.

What the Fed Aims to Achieve with QE2

Essentially, the Fed sees its policy as having three mutually-reinforcing goals.

First, they want to create inflation or raise inflationary expectations in the hope that this will get consumer spending revved up again and lead to a revival of the housing market. The idea here is that if households expect inflation to rise they will go out and buy now when prices are lower.

Second, the Fed wants to lower long-term interest rates. The rationale here is that lower interest rates will also spur borrowing and spending by households and corporations. As consumer spending picks up, companies will be able to expand their production facilities to meet the increased demand by borrowing at low rates.

Third, by hitting the electronic presses, the Fed knows that another possible consequence of flooding the economy with a further wall of cash is that the US dollar will weaken. But, this would be a welcome side-effect as it would give US exporters a big leg-up in international markets by boosting their competitiveness.

Thus, the Fed's latest round of quantitative easing is aimed at pulling the US economy from its current quagmire by shoring up domestic demand and boosting its international competitiveness. The hope is that a combination of lower borrowing costs, rising inflationary expectations, and a falling dollar will jolt the economy and put it onto a higher growth trajectory.

Fed's Intentions Laudable but Misplaced

While the Fed's intentions are laudable, what is puzzling about the their recent move is that they are acting as if the US economy is again in the same extreme distress as it was during the darkest quarters of the 2008-2009 recession. During that period, the economy was in a tail spin, credit markets were frozen, hundreds of thousands of jobs were being lost every month, house prices were plunging, and stock markets were gyrating wildly. Not surprisingly, the nation was gripped in fear during those months.

The Fed's and the US government's quick and multi-faceted policy response during those dark days was entirely appropriate. The combination of near-zero interest rates, the purchase of $1.7 trillion in government and agency bonds via the first round of quantitative easing, the $787 billion stimulus package, and the federal government's bailout of a number of collapsing major financial institutions all helped to shore up the US economy and rescue it from falling into an even deeper hole.

US Economy on Slow Mend

Although the Fed's aggressive policy response during the Great Recession was fitting, one has to question the validity of its latest initiative. While media headlines typically tend to convey the sense that the US economy is still struggling to break free from the grip of recession, a quick look at the facts clearly shows that the economy is actually on the mend – albeit slowly.

Since the recession's official end in the second quarter of 2009, the nation's GDP has been growing for five consecutive quarters, consumer spending and business investment are also rising, and exports have been climbing helped by the weaker dollar.

On the inflation front, after dipping into deflationary territory in 2009 when the consumer price index fell by 0.3%, overall prices have been in positive territory throughout this year. The latest numbers show that the CPI increased by 1.2% in October from a year earlier.

On the job front, the private sector has created over 1.1 million jobs since December 2009 and the total number of people collecting unemployment benefits fell by 86,000 to 4.73 million in October, the lowest level since November 2008.

However, given the slow pace of the recovery, the unemployment rate remains stubbornly stuck at 9.6% showing that the recession is not over for a large number of Americans. Still, the key point is that the US economy is gradually getting better and the recovery is becoming more secure.

US Interest Rates Already at Near Historic Lows

What is particularly perplexing about the Fed's recent move is its intention to drive down interest rates – when they are already at rock-bottom levels – by injecting yet more hundreds of billions of cash into the financial system. In fact, since hitting their 2010 highs on April 5, yields on US Treasuries have plunged across the board.

Between April's peak and the November 3 QE2 announcement, interest rate declines have ranged from a low of 5 basis points (or 0.5 percentage point) on 3-month T-bills to a high of 164 basis points for 5-year notes. In percentage terms, the yields on 3-year bonds have dropped the most, falling by 72.3% (see Table 1)

TABLE 1
TRACKING US TREASURY YIELDS
(Yields were already at near-record lows before the Fed's QE2 announcement)

	3 Nov 2010[1/]	5 Apr 2010[2/]	Basis Point Change	% Change
3-months	0.13	0.18	-5 bps	-27.8%
1-Year	0.22	0.48	-26 bps	-54.2%
3-Year	0.49	1.77	-128 bps	-72.3%
5-Year	1.11	2.75	-164 bps	-59.6%
7-Year	1.85	3.46	-161 bps	-46.5%
10-Year	2.67	4.01	-134 bps	-33.4%
30-Year	4.09	4.85	-76 bps	-15.7%

[1/] QE2 announced, [2/] 2010 highs in yields
Source: US Department of the Treasury

Mortgages rates are also at their lowest levels in decades. The average rate for 30-year fixed loans has fallen from 5.1% in April to 4.23% now and similarly the average rate on 15-year fixed loans is now 3.63%, down from April's 4.42%.

QE2 Raising Alarm Bells

Following the Fed's QE2 policy announcement on November 3, Treasury bond yields initially fell. But, as investors began to digest the implications of the Fed's move, alarm bells started to ring warning that their actions could spark an outburst of inflation, drive down the dollar, and destabilize the global economy.

As a consequence, investors have started to sell off US government bonds and yields have been rising across the board since November 5, a mere two days after the announcement. For example, the 5-year yield jumped 40 basis points to 1.51% on November 15 from its 1.11% low on November 3 and the 10-year yield increased by 25 basis points to 2.92% during this time frame.

The Fed's plan to stimulate demand with a further injection of liquidity is unlikely to achieve very much given that US corporations are already sitting on more than a $1 trillion of cash which is more than enough to meet any increased demand. Moreover, with interest rates already at rock-bottom levels, it really is a stretch to believe that a reduction of a few more basis points will spur American consumers to borrow and ramp-up their spending to boost economic activity.

Indeed, despite historically low interest rates, the recovery is likely to continue to plod along for an extended period of time for one simple reason – the deleveraging process that is underway still has quite a long way to go before household balance sheets are in better shape.

The Bottom Line

Given that the US economy is growing and, therefore, there really is no pressing need for another heavy dose of monetary stimulus, one has to question the wisdom of the Fed's actions. The economy is moving forward, not backward as the Fed's actions would lead one to believe. Moreover, companies are refinancing their debt at record-low rates and improving their balance sheets, and households are beginning to get their finances in order.

Unless the latest round of quantitative easing produces clear and significant sustainable benefits, the Fed is at risk of debasing its very credibility. It needs to step back and reflect on the consequences of its policy, one that could be construed as nothing more than monetary adventurism.

As the great Scottish poet Rabbie Burns aptly wrote in 'To a Mouse' one of his classic poems *'...the best-laid schemes o' Mice an' Men, Gang aft a-gley'*, it would seem that the best laid plans of the Federal Reserve are also showing signs of going sadly awry.

23

Germany Emerging As Europe's Powerhouse

(December 2010)

The German economy, despite being the only advanced Western nation not to experience a housing bubble and having a competitive economy and its government finances in good shape, was nevertheless hit hard by the recent global financial crisis.

From the beginning of the downturn in the first quarter of 2008 to the recession's trough a year later in the first quarter of 2009, the country experienced its deepest slump in the post-war period as GDP declined by 6.6%. In contrast, during these four quarters, output fell by 3.8% in the United States, 3.2% in France, and 5.5% in the United Kingdom.

What led to Germany's steep slide was the collapse in world trade. With exports accounting for more than 50% of the country's output, the drying up of foreign markets dealt a massive blow to the economy. As global credit markets froze cutting off the avenue to trade financing, foreign demand for German products plunged and exports fell by 17%. This had a knock-on effect on investment spending which also went into a tailspin falling by 12.3%.

The government's swift implementation of its Kurzarbeit or 'short work' policy which, by providing wage subsidies, encourages employers to reduce hours rather than lay-off workers during economic downturns, limited the recessions impact on unemployment. This in turn helped to shore up consumer spending which dipped by only 0.2% during the recession. Government expenditures, which accounts for about a fifth of total output, was the only major component of GDP that increased during the recession (see Table 1).

TABLE 1
GERMANY'S ECONOMY IN RECESSION & RECOVERY
(Cumulative % change)

	Recession	Recovery
	Q1 2008 – Q1 2009	Q1 2009 – Q3 2010
Consumer Spending	-0.2	+0.5
Gov't Expenditure	+3.2	+3.1
Capital Investment	-12.3	+10.6
Exports	-17.0	+18.9
Imports	-8.3	+14.3
REAL GDP	-6.6	+5.2

Source: OECD

German Recovery Faster than Expected

Just as the collapse in world trade buckled the German economy, its resurgence, driven mainly by the strong growth in emerging markets, has quickly put the country back on its growth track. Since exiting from recession in the first quarter of 2009, the economy has been growing for six consecutive quarters with all major components of GDP contributing to the recovery.

From the start of the recovery to the third quarter of this year, GDP has increased by 5.2% and the level of output is now just 1.8 percentage points shy of the pre-recession peak level reached in the first quarter of 2008. Moreover, the pre-recession output level is now expected to be attained by 2011, two years ahead of the earlier projections made this spring by economic forecasters including the IMF and the OECD.

Recent survey data all point to a recovery that is not only broadening but is also gaining in strength. Business confidence is rebounding strongly and, according to the widely followed Ifo survey by the Munich-based research institute, it reached its highest level this December since the country's reunification in 1990.

Moreover, with unemployment falling and wages starting to rise, German consumers are also in an upbeat mood. The latest survey by Germany's GfK institute shows that consumer confidence levels are now at a three-year high which bodes well for the retail sector.

The economy is currently enjoying a virtuous cycle where output is rising, jobs are being created, and unemployment is falling. Indeed, while the German economy has recovered 98.2% of the output that was lost during the recession, the labor market has recouped all of the job losses that occurred during the downturn.

Employment has been rising steadily over the past several months and it reached a new record high of 41 million in October. This has helped push the unemployment rate down and it is now below its pre-crisis level. Moreover, the number of Germans that are unemployed has dropped below the 3 million mark for the first time in 18 years.

Growth Now Broad Based

In the first few quarters coming out of the recession, exports and inventories were the major contributors to economic growth. But the recovery is becoming more balanced and domestic demand is now contributing more than half of the country's growth in GDP. Consumer spending is rising and investment spending is rebounding strongly and is expected to be a key driver of the recovery.

With domestic demand picking up, Germany's economic outlook continues to brighten. Going forward, as the economy rebalances and domestic demand becomes the dominant driver of the economy, growth in imports is expected to outpace exports and as a result the contribution of net exports to GDP growth is expected to diminish.

The pickup in domestic demand is also helping to reduce the country's current account surplus. As a percent of GDP, Germany's current account surplus has fallen from a record high of 7.7% in 2007 to 4.9% in 2009 and is projected to fall further to around 4% by 2012.

Germany's GDP is projected to rise by 3.7% in 2010 and the economic expansion is expected to continue in 2011 but at a slower pace given the anticipated slowdown in the country's major trading partners and the reduction in government spending. Nevertheless, growth is expected to be well above Germany's trend rate of growth which is estimated to be between 1% and 1.5%.

In addition, Germany's government finances are also in good shape and are set to get better. As fiscal policy starts to shift from stimulus to consolidation the budget deficit is expected to drop to by a full percentage point to 2.7% in 2011 and decline further to around 2% in 2012.

The Bottom Line

With its economy going from strength to strength, Germany is rapidly emerging as Europe's powerhouse. Indeed, some would say that it is in the process of reclaiming its traditional role. As it becomes the benchmark by which other European economies will be judged, watch for an increasingly assertive Germany. It will be in the driver's seat when it comes to calling the shots on the economic policy front in the Eurozone. How this will be received by the other member countries remains to be seen

PART 3

The Year 2011

24
Canada's Economy Faces Challenges

(January 2011)

Compared to most other advanced economies, Canada has weathered the Great Recession of 2008 and 2009 relatively well. From peak to trough, the country's real GDP declined by 3.4% whereas it fell by 4.4% in the United States, 5.3% in the Eurozone, and 6.5% in the United Kingdom. The increase in the country's unemployment rate from 6% to a high of 8.7%, while wrenching, was also considerably less severe than that of the US. South of the border, the jobless rate more than doubled; it jumped from under 5% before the onset of the downturn in the fourth quarter of 2007 to a high of 10.1%.

There are a number of reasons why Canada's economy held up comparatively well. While the trade sector and business investment nosedived, consumption – shored up by automatic stabilizers and the government's stimulus measures – remained resilient. Moreover, unlike the United States, the country avoided the meltdown in the housing market that continues to plague the US economy.

In addition, strong demand for the country's natural resources, especially from emerging economies including China and India, helped to mitigate the general collapse in global demand.

Economy Recovers Lost Ground

Since exiting the recession in the second quarter of 2009, Canada's economy has posted positive growth for five straight quarters. Indeed, the latest national accounts data, which is for the third quarter of 2010, shows that at the aggregate level the economy has recovered all the ground it lost during the recession.

Similarly, on the employment front, the economy has also recouped all the jobs that were lost during the recession, although the number of full-time jobs is still about 85,000 below its previous peak of 14 million achieved back in February 2008.

On the surface, it does appear that Canada's economy is quite robust and healthy. But, digging deeper into the data and taking a broader perspective, one finds that all is not quite what it appears on the surface.

An Increasingly Unbalanced Economy

What stands out from an examination of the macroeconomic data is that the Canadian economy is increasingly becoming disproportionately dependent on both consumer and government spending. As a share of the economy, consumer spending is now at a record high and accounts for 64% of Canada's GDP (see Table 1). In fact, the share has been rising every single year since 2000 and currently is up a whopping eight percentage points from a decade ago.

Total government spending has also increased its share of the economy increasing from about 21% in 2000 to over 25% now. In the first half of the last decade, government spending, as a proportion of GDP, ranged between 21% and 22%. It has since climbed but this is primarily due to the effects of the recession. However, as policy makers start to curtail public sector spending and rein in the budget deficit, which now stands at over 5% of GDP (at both federal and provincial levels), the government's share as a percent of the nation's output should start to nudge down.

TABLE 1
THE CHANGING STRUCTURE OF CANADA'S ECONOMY
(Percent Distribution of Real GDP by Major Component)

	2000	2005	2010
Consumer Spending	56%	58%	64%
Government Expenditures	21%	22%	25%
Business Investment	17%	20%	18%
Exports	44%	40%	33%
Imports	(40%)	(41%)	(43%)
Net Exports*	+4%	-1%	-10%

* Exports less Imports. Source: Statistics Canada

Business investment, which typically tends to be more sensitive to the economic cycle, has more or less remained the same over the past decade averaging about 18%. While investment spending, particularly in machinery and equipment, has picked up since the start of the recovery in mid-2009, it is still about 10% below its pre-recession peak.

Export Sector Falters but Imports Hold Their Ground

What is particularly startling, however, is the sharp decline that has taken place in Canada's export sector over the past decade. As a percentage of GDP, the share of exports has plunged eleven percentage points, falling from 44% in 2000 to 33% today. Furthermore, even before the recession hit, it is interesting to note that this ratio declined by four points between 2000 and 2005, a period when growth in the United States and the world economy was quite robust. The onset of the global recession basically accelerated the relative decline in Canada's exports sector.

While the share of exports has been dwindling, the share of imports, particularly over the past five years, has been rising (see Table 1). After dropping sharply during the recession, imports have increased by nearly 20% and are now only about 3% below the pre-recession peak. In contrast, exports have risen by 8% but are still nearly 13% below the previous peak. As a result of these divergent trends, the gap between exports and imports has widened significantly. As a percent of GDP, imports now exceed exports by a staggering record-breaking ten percentage points.

Furthermore, imports have now exceeded exports every single quarter for the past eighteen quarters and what is more worrisome, the gap has been widening sharply. From the start of the recovery in the second quarter of 2009 to the third quarter of 2010, the negative gap between real exports and imports (net exports) has surged from $75 billion to $137 billion.

Myriad Risks Loom on the Horizon

While Canada's economy has recovered all the ground it lost during the recession of 2008 and 2009, it remains vulnerable to a number of factors which could stymie its future growth prospects. In particular, a frothy housing market where prices are overextended, record high household debt, a surging Canadian dollar, and a weak US economy are all factors that could restrain GDP growth.

Fuelled by cheap credit and to all intents and purposes dismissive of the recession, Canadians have been loading up on debt in record amounts. The latest figures from the Bank of Canada show that total household debt reached $1.47 trillion in September 2010, up 23% from $1.19 trillion in December 2007. As a percent of disposable income, Canadian household debt is now at 148% – a record high.

Canadian policy makers are well aware that a high household debt level puts the economy in a vulnerable position. Indeed, the Governor of the Bank of Canada has been exhorting Canadians about the dangers of carrying high debt levels. The worry is that a sudden spike in interest rates or a correction in house prices that are also in record territory, could force Canadian consumers to rein in their spending and thereby jeopardize the recovery.

Bank of Canada Faces Difficult Balancing Act

Despite recognizing the dangers of excessive household debt, the Bank of Canada faces a dilemma. The standard remedy to curb credit growth is to raise interest rates. But higher rates are likely to add further fuel to the Canadian dollar which would intensify the woes of the export sector.

As a major commodity currency in high demand and with its economy in relatively better shape than other G7 nations, the Canadian dollar has been on a tear over the past several months. Since December 2008, the *loonie* has gained 22% against the US dollar, 25% against the Euro, 17% against the British pound, and 12% against the Japanese yen.

On the other hand, by leaving interest rates at abnormally low levels when the economy is growing only leads to further asset price distortions particularly on the housing front. Canada's housing market has been on a bull run for a decade and, although the government has tightened mortgage rules over the past few months, it remains vulnerable to a correction.

Indeed, according to the OECD's data on house price ratios, Canada's housing market is clearly overstretched. The latest figures, which are for 2009, show that the country's home price-to-rent ratio is 46.4% above its long term average and the price-to-income ratio is 27.7% higher than its historical norm.

In stark contrast, the price-to-rent ratio has been falling in the US since the collapse of the housing bubble in 2006 and is now only 13% above its long-term average. Moreover, its price-to-income ratio has now dropped – by one percentage point – below its historical average. Both these trends strongly suggest that the steep downturn in the US housing market is coming to an end. Unfortunately, in Canada's case, the opposite may be about to begin.

Near-Term Outlook

Canada's economy is expected to keep growing in 2011 but with the unwinding of fiscal stimulus and the need to rein in the budget deficit, growth could start to fade. Moreover, Canada cannot count on the US economy to boost demand for its exports.

The US is still struggling with its recovery and faced with high unemployment, a still weak housing market, and the pressing need to rein-in its mountainous budget deficit that exceeds $1 trillion, at best growth is likely to remain below trend for quite some time yet. While growth over the past year has been strong in emerging economies, the region accounts for less than 10% of Canadian exports.

The Bottom Line

As noted above, over the past decade the composition of demand in the Canadian economy has become quite skewed. In order to secure a stronger and more sustainable recovery over the longer term, the economy needs to rebalance its underlying structure of demand by moving more toward investment and exports and relying less on consumption that is being increasingly driven by debt.

It's easy to say, but a challenge to achieve.

25

Inflation And The G7 Economies

(February 2011)

As the global economy rebounds from the financial crisis and gathers pace, inflation is beginning to kick in and is causing rumblings of concern. Fuelled by the surge in food and energy prices, inflation is now on the march and is rising world-wide. In the emerging economies, the inflation rate in China increased to 4.6% in December and was up by 5.9% in Brazil, 8.8% in Russia and 9.7% in India.

In the G7 countries higher commodity prices are also pushing up the rate of inflation and, as one can see from Table 1, the headline rate of inflation has risen in all of the G7 economies. Across the group as a whole, inflation rose by 1.4% in 2010 up from –0.1% in 2009. Excluding Japan, where consumer prices fell again last year, the rise in inflation ranged from a low of 1.1% in Germany to a high of 3.2% in the United Kingdom.

While the rise in headline inflation is largely due to higher food and fuel prices, the core rate (which strips out these components) remains quiescent in the advanced economies. Indeed, core inflation has actually been trending down and it fell in all G7 countries last year. The only exception was Britain where the core rate moved higher (see Table 1).

The dip in underlying inflation suggests that the price surge in commodities has not, at least as yet, spilled over into other sectors of the economy.

With several of the big emerging economies overheating, central banks in China, India, Indonesia and Thailand are all raising interest rates in an attempt to curb growth and rein in escalating prices. But, while tightening monetary policy and raising interest rates to combat inflation is the correct policy response in these countries, the situation in the advanced economies is less clear-cut.

TABLE 1						
INFLATION TRENDS IN THE G7 ECONOMIES						
	Headline[1]			Core[2]		
	2009	2010	Trend	2009	2010	Trend
Canada	0.3	1.7	Up	1.7	1.3	Down
France	0.1	1.5	Up	1.7	0.9	Down
Germany	0.4	1.1	Up	1.3	0.7	Down
Italy	0.8	1.5	Up	1.7	1.6	Down
Japan	-1.4	-0.8	Up	-0.6	-1.2	Down
Britain	2.2	3.2	Up	1.8	2.9	Up
USA	-0.4	1.6	Up	1.7	1.0	Down
G7	-0.1	1.4	Up	1.4	0.9	Down

[1] Consumer Prices – all items; [2] Consumer Prices – all items excluding food and energy. Source: OECD

Although inflation is also picking up in the G7 countries, the recovery that has been underway since mid-2009 is still quite tentative. Factories are not operating at full capacity, job creation is anemic, and the unemployment rate remains stuck at high levels and is well above pre-financial crisis levels. Moreover, loan-demand by households and companies continues to remain weak.

Central Bank Dilemma

Nevertheless, despite the sub-par recovery, the rise in the headline rate of inflation does present a quandary for western central banks who face a tricky choice – either keep interest rates low and support the economic recovery, or raise them to try and stop inflation from getting a foothold. While central banks focus primarily on core inflation in determining the direction of monetary policy, the concern is that rising commodity prices will sooner or later ripple through the economy and ultimately affect the core rate.

However, it's the headline rate of inflation, not the core rate, that plays a major role in influencing the wage demands of consumers. And here, recent survey data shows that inflationary expectations are starting to edge up in several of the G7 countries raising the prospect that, unless checked, these expectations could become entrenched. History shows that once inflation takes hold it is difficult and costly to uproot.

A key question in investors' minds is, given the rise in the headline rate of inflation, will central banks be forced to tighten monetary policy sooner than expected? Given the slack in the G7 economies, the consensus view has been that the Federal Reserve and the European Central Bank would leave rates alone for much of 2011. But the concern now is that higher commodity and energy prices could push up the prices of other goods and services. This in turn could prompt workers to demand higher wages and, unless checked, inflation could become embedded.

But how realistic is this viewpoint? A look at the data shows that, with the exception of Britain, inflation in the G7 remains contained. There is still a significant amount of unused capacity in the economy, demand remains weak, and the high jobless rate is keeping a lid on wage increases.

Given these factors it is hard to see a return of high inflation. Certainly, the large amount of monetary stimulus that has been pumped into the G7 economies is of concern, but at this stage of the economic recovery the risk of inflation taking off is quite remote.

Time for Central Banks to Start Normalizing Interest Rates

Still, that does not mean that policy makers should maintain the status quo when it comes to interest rates. Certainly, the slashing of interest rates to record-lows, following the collapse of Lehman brothers in the fall of 2008, was the correct policy response by the major central banks. Moreover, together with the governments' fiscal stimulus measures, it did the job in that it prevented the global recession from turning into another Great Depression.

But, with the recovery advancing in all the G7 countries, one really has to question the need for keeping rates at these record-low 'emergency' levels. Monetary policy in the major advanced economies remains very loose and real interest rates (nominal rates minus inflation) are negative in all the G7 economies and range from –3.2% in the United Kingdom to –0.8% in France.

To prevent a wage-price spiral from taking hold as well as putting a stop to new speculative bubbles from forming, central banks need to start the process of normalizing monetary policy by gradually increasing interest rates, albeit in small increments, in line with inflation. By keeping policy rates unchanged while inflation rises essentially amounts to a further easing of monetary policy and potentially sets the stage for yet another round of asset price bubbles.

So far, Canada is the only G7 country that has increased interest rates. The central bank hiked rates three times in 0.25% moves to 1% over a four month period in 2010, but since last September it too has kept rates on hold.

The Bottom Line

In its latest forecast, the IMF projects that GDP growth for the G7 group of countries will average 1.9% in 2011 ranging from a low of 1% in Italy to a high of 3.0% in the United States.

With the recovery proceeding in all the major economies, the time has come for the central banks in the advanced countries to start laying the groundwork for the gradual return to a more normal monetary policy framework and one that is geared to maintaining price stability over the medium term.

26

The Fed's Zero-Interest Rate Policy

— *It's Time for a Change* —

(March 2011)

During the global recession the US Federal Reserve, concerned about the extent of the economic downturn and the growing threat of deflation, moved aggressively to contain the situation. They did this by repeatedly slashing interest rates close to zero in order to stimulate demand.

But, despite these sharp cuts in interest rates the slump gathered momentum and continued to deepen especially in late 2008 and early 2009 when real GDP plunged by 6.9% in the final quarter of 2008 and fell by another 4.9% in the first three months of 2009.

Clearly worried that the conventional monetary policy response of sharply lower interest rates was not having the desired outcome and the fact interest rates could not be cut further, the Fed opted to pursue a policy of quantitative easing. By electronically printing money in order to buy government bonds and other longer term securities, the goal of quantitative easing was to lower long-term interest rates (bond prices and rates move in opposite directions) and thereby lend further support to the economy.

Quantitative Easing and the Fed's Surging Balance Sheet

In the first round of quantitative easing (QE1), which ran from December 2008 through March 2010, the Federal Reserve purchased about $1.7 trillion in longer-term Treasury, agency, and agency mortgage-backed securities. Bond yields initially fell in the first month of the program but then started to drift up. For example, the yield on 5-year US government bonds fell by 77 basis points to 1.52% in December 2008 but by the time the first round was over, yields had risen to 2.43%. Similarly, yields on 10-year bonds followed an analogous path.

QE1 was followed by the second round of quantitative easing (QE2) in early November 2010 when the Fed announced a plan to purchase an additional $600 billion in longer-term Treasury securities by June 2011. However, this time around bond yields have been rising since the start of the program effectively blunting the Fed's intended goal of seeking to push long-term rates lower.

Nevertheless, as a result of the Fed's ultra-loose monetary policy, its balance sheet has soared to over $2.6 trillion and it is set to reach almost $3 trillion by the middle of this year when the second round of quantitative easing is completed.

Pace of Inflation Quickening

Despite the Federal Reserve's claims that the effects of increased commodity costs on inflation are expected to be transitory, the latest data shows that the pace of inflation is actually quickening. US consumer prices jumped to 2.1% in February which was almost double the 1.1% rate that was recorded just four months earlier in November.

Similarly, core inflation, which strips out the food and energy components from the overall index, is also rising. The core index increased to 1.1% in February after reaching a record low of 0.6% in October.

In addition, the latest consumer survey from Thomson Reuters and the University of Michigan shows that inflation expectations are mounting. The survey's one-year inflation expectation index jumped to 4.6% in March, up from 3.4% a month earlier. Indeed, as recently as last September, US consumers were expecting inflation to rise by 2.2% annually.

As global food and energy prices soar, inflation is rising worldwide and without concerted policy action it is unlikely that it will be contained. History shows that once inflation takes hold, it is fiendishly difficult to rein in. Although the consensus view is that the Fed's first rate hike is not expected until the second half of next year at the earliest, it is nevertheless worth remembering that inflation is capable of accelerating – even during recessionary periods.

For example, during the 1973-75 US slump, one that was induced in part by surging oil prices, GDP declined and the unemployment rate shot up. Nonetheless, the rate of inflation sped up. US consumer prices were increasing at 8.4% annually when the recession started in the fourth quarter of 1973 and by the time it was over in the first quarter of 1975, prices were rising by 11.1% annually.

Fed Falling Behind the Curve

It is interesting to note that, unlike the US Federal Reserve, the European Central Bank (ECB) is getting quite concerned about rising inflation. Indeed, policy makers at the ECB, after voicing their determination to keep inflationary expectations from taking hold, have done an about face in its monetary policy stance.

As a result, it is now virtually a certainty that the ECB will raise its benchmark rate, which is currently at a record low 1%, as early as April. While this will signal the start of a tightening cycle, given the tenuous state of the recovery in several Eurozone economies, it is unlikely that the initial rate increase will be quickly followed by a series of further hikes.

The Fed, however, has indicated that it will not go down this route. Fed Chairman Bernanke recently stressed again that unemployment in the US remains too high and while the recovery is on firmer footing monetary policy needs to remain supportive.

However, with inflationary pressures rising, there is the danger that the Fed is now falling behind the curve in its execution of monetary policy. Ironically, the Fed was first off the mark and ahead of the curve among the world's major central banks during the recession.

The Bottom Line

The Fed will face a challenge when it comes to unwinding its large balance sheet and, with little historical precedent to go by, it will likely be a tricky process. With the economy recovering – in fact the loss of output has completely regained its pre-recession level – and with inflationary expectations starting to rise briskly, the Fed needs to start the process of withdrawing the trillions of dollars it has been printing. Moreover, this should be done sooner rather than later. If not, it runs the risk of letting inflation rip loose.

The time has come for the Fed to change the stance of its monetary policy as the crisis-induced super accommodative policy it has been pursuing is no longer warranted. It needs to move away from being ultra-loose to being just loose.

27

Recession & Recovery In The Eurozone
— A Tale of Diverging Paths —

(April 2011)

The European economy is slowly emerging from the Great Recession which took a sharp toll on GDP and sent the unemployment rate in the Eurozone soaring from 7% in 2007 to almost 10% in 2010. Although all the member states were hit by the crisis, its impact on the individual countries differed significantly.

Similarly, even though economic growth has resumed in virtually all the Eurozone countries, the path to recovery varies widely. In this paper we evaluate the damage the recession inflicted on the Eurozone economies and weigh the strength of the recovery in the member countries.

Recession's Impact on GDP

Table 1 highlights the impact of the global recession on GDP for the Eurozone economies. As you can see, the impact of the slump on the overall level of output varied considerably across the individual member countries hitting some nations much harder than others. The length of the recession also varied significantly and ranged from the relatively short 3 quarters in the case of Belgium to a staggering 12 quarters, or 36 months, for Ireland.

All three of Europe's biggest economies entered into recession at the same time but the slump lasted for four quarters in Germany and France and five quarters in Italy. In terms of the impact on GDP, the more open economies of Germany and Italy suffered steeper declines in output (–6.6% and –7.0% respectively) than France where real GDP fell by 3.9%.

TABLE 1
DEPTH & DURATION OF THE RECESSION IN EUROZONE ECONOMIES

	GDP Peak Quarter	GDP Trough Quarter	Quarters in Recession[1]	Cumulative Decline in GDP[2]
Ireland	Q4 07	Q4 10	12	-14.6%
Finland	Q2 08	Q2 09	4	-10.2%
Greece	Q3 08	Q4 10	9	-8.9%
Italy	Q1 08	Q2 09	5	-7.0%
Germany	Q1 08	Q1 09	4	-6.6%
Austria	Q2 08	Q2 09	4	-5.4%
Netherlands	Q1 08	Q2 09	5	-5.3%
Spain	Q1 08	Q4 09	7	-4.9%
Belgium	Q2 08	Q1 09	3	-4.2%
Portugal	Q4 07	Q1 09	5	-3.9%
France	Q1 08	Q1 09	4	-3.9%

[1] Number of negative quarters; [2] From peak to trough; Source: OECD

Among the PIGS, Ireland's recession has been particularly severe. Between the fourth quarter of 2007 and the fourth quarter of 2010, real GDP fell by a cumulative 14.6%. Portugal's economy also fell into recession at the same time as Ireland's but its downturn was considerably shorter. Portugal's slump lasted five quarters and the economy contracted by 3.9% during this period.

Interestingly, Greece's economy, after peaking in the third quarter of 2008, was the last Eurozone economy to topple into recession. But, instead of following the old adage of last in, first out, Greece will likely be the last of the Eurozone countries to climb out of recession. The economy has now been contracting for over two years and GDP has fallen by 8.9%. Finally, Spain's recession persisted for seven quarters during the course of which GDP declined by 4.9%.

In the smaller Eurozone members, Finland's export-dependent economy suffered a steep recession as the financial crisis and the collapse in global trade sent its GDP plunging by 10.2%. The economies of Austria, Belgium and the Netherlands, also contracted but the decline in output was less severe and ranged from 4.2% to 5.4%.

An Uneven Recovery Path

With the exception of Ireland and Greece, most Eurozone economies are recovering, but growth remains highly uneven. As in the case of the recession, the pace and duration of the recovery has also varied considerably (see Table 2). Among the Big 3, the economies of Germany and France resumed growing in the second quarter of 2009 and were followed by Italy a quarter later. Germany has been the top performer in this group and after seven straight quarters of growth the country's GDP is up 5.5% from the bottom.

In the PIGS, growth has resumed in Portugal and Spain but recessionary forces continue to grip the economies of Ireland and Greece. Portugal's economy pulled out of recession in the second quarter of 2009, but its recovery has been weak. Similarly, although Spain's economy has been in recovery mode for the past four quarters, the revival in growth has also been very anemic.

Returning to Pre-Recession Levels – A Long Slog Lies Ahead

Despite the return to growth, the level of output still remains below pre-crisis levels in all of the Eurozone economies (Table 2, column 5).

Amongst the largest economies, Germany's GDP is currently 1.4% below its pre-recession peak and the comparable figure for France is 1.6%. However, given Italy's feeble recovery, its economic output is still 5.2% below its pre-recession level and has the most ground to make up, not only in relation to France and Germany, but also amongst all the Eurozone countries.

TABLE 2
THE UNEVEN ROAD TO RECOVERY
IN EUROZONE ECONOMIES

	Start of Recovery	Quarters in Recovery	Cumulative Rise in GDP[1]	% From Previous Peak
Belgium	Q2 09	7	3.7%	-0.7%
Germany	Q2 09	7	5.5%	-1.4%
Austria	Q3 09	6	4.2%	-1.4%
France	Q2 09	7	2.4%	-1.6%
Portugal	Q2 09	7	2.0%	-2.0%
Netherlands	Q3 09	6	3.4%	-2.1%
Spain	Q1 10	4	0.6%	-4.3%
Finland	Q3 09	6	6.3%	-4.5%
Italy	Q3 09	6	1.9%	-5.2%
Greece	In Recession – GDP continuing to contract			-8.9%
Ireland	In Recession – GDP continuing to contract			-14.6%

[1]/From start of recovery to 2010 Q4; Data Source: OECD

The anticipated timeline for a full recovery in the big 3 also varies significantly. For example, assuming Germany maintains its average quarterly growth rate of 0.8% that it has recorded during the past seven quarters, it should return to its pre-recession level by the third quarter of this year. France's economy is expected to return to its former peak by the first-half of 2012 but more than likely it will take at least another three years before Italy's economy recoups all of the lost output it lost in the recession.

When it comes to the PIGS, the timeline for a full recovery is considerably bleaker. For example, Spain's economy has posted quarterly growth of only 0.15% during the past year and at this rate it won't be until 2017 before Spain's GDP gets back to its pre-recession peak level. Paradoxically, while the focus of attention in recent weeks has been on the debt crisis that is plaguing Portugal, given that the country's GDP is only 2% below its previous peak, it may be the first economy in this group to close the gap.

However, this is certainly not the case with Ireland and Greece. With their economies still mired in recession – some would say depression – it will take several years, perhaps even a decade or more, before the level of output in these two countries returns to their respective 2007 pre-crisis levels.

The Bottom Line

There is little question that the recovery remains weak in most Eurozone countries and, as the risks to the outlook continue to multiply, the ability of policy makers to speed-up growth is unfortunately virtually non-existent. Indeed, with an already lacklustre recovery, there is a real danger that as the austerity measures, including tax hikes and sharp cuts to public spending, start to bite this will further push down growth rates.

Moreover, when one adds in the mix of rising interest rates, volatile oil prices, and an appreciating euro, then it becomes glaringly clear that any near-term prospect of a resurgence in growth vanishes. Given this toxic environment, several of the Eurozone economies including Spain, Portugal, and Italy could find themselves sliding back into recession and dragging down the region. Not a happy picture.

28

The Great Recession's Impact On Government Finances In The G7

(June 2011)

The financial crisis and the ensuing global recession have dealt a crippling blow to public sector finances in the G7 and from which all member countries have yet to recover. As GDP contracted and the unemployment rate surged government revenues fell and spending jumped in all the major economies but with wide variations.

In terms of revenues, Japan's public sector was the hardest hit – revenues fell by 12.3% between 2007 and 2009 – and was followed by the United States (-8.6%) and Canada (-6.2%). Interestingly, the recession's impact on government revenues was more muted in Europe and ranged from a 3% drop in the case of the United Kingdom to a 1.4% decline in Italy. Germany was the only G7 nation where revenues held steady during the downturn (see Table 1).

On the spending side, as the recession deepened, automatic stabilizers such as unemployment compensation and welfare payments kicked in which helped to mitigate the downturn.

Furthermore, to stop their economies from plunging into an even deeper hole, G7 policy-makers introduced a variety of fiscal stimulus measures ranging from programs on infrastructure spending, to tax breaks, to auto buying incentives.

Unsurprisingly, the combination of automatic stabilizers and fiscal stimulus measures sharply ramped-up government spending. Between 2007 and 2009, the increase in public sector spending ranged from a low of 6.4% in Italy to a high of 19.4% in the United States.

TABLE 1
THE GREAT RECESSION'S IMPACT ON GOVERNMENT REVENUES & EXPENDITURES IN G7 ECONOMIES
(2007 – 2009)

	Total Revenues	Total Expenditures
Canada	-6.2%	+11.7%
France	-1.7%	+7.7%
Germany	0.0%	+7.5%
Italy	-1.4%	+6.4%
Japan	-12.3%	+9.6%
United Kingdom	-3.0%	+15.9%
United States	-8.6%	+19.4%

Source: IMF

As a result of these sharp divergences in government revenues and expenditures, the fiscal deficit has risen in all the major economies. Before the onset of the global recession, only Canada and Germany were running surpluses but by 2009 all the G7 economies were in the red. The fiscal deficit in 2009, as a share of GDP, ranged from a low of 3% in Germany to a high of 12.7% in the United States.

A Look Ahead at Government Finances

According to the International Monetary Fund's latest medium-term projections, over the 2010-2015 period government revenues are projected to increase at a faster pace than the overall growth in the economy in all of the G7 countries. On the other hand, government spending is projected to grow at a considerably slower pace in all cases (see Table 2).

For example, government revenues in the United States are forecast to increase by 40.8% by 2015 while spending is projected to rise by 21%. In the United Kingdom revenues flowing into government coffers is projected to rise by 33.7%, triple the rate of growth in spending. Even in Japan, notwithstanding the fact that the country's population is declining and the labour force is shrinking, revenues are nevertheless forecasted to outstrip spending.

TABLE 2
GOVERNMENT REVENUES & EXPENDITURES PROJECTIONS FOR G7 ECONOMIES
(2010 – 2015)

	Total Revenues	Total Expenditures
Canada	34.0%	17.7%
France	23.4%	11.2%
Germany	16.7%	8.7%
Italy	15.3%	12.0%
Japan	14.1%	6.7%
United Kingdom	33.7%	10.3%
United States	40.8%	21.0%

Source: IMF

Should the gap between revenues and spending which is expected to widen significantly over the next five years materialize, this would sharply reduce the budget deficits of the major economies. Still, by 2015, only the budgets in Canada and Germany are expected to be in balance. In the rest of the group the deficit, as a percent of GDP, is projected to be 7.4% in Japan, 5.5% in the United States, 3.1% in Italy, 2.3% in the United Kingdom, and 2.2% in France. The question, of course, is how realistic are these projections?

IMF Growth Forecasts Overly Optimistic

The GDP forecasts undoubtedly are the wild card and here the IMF clearly has an optimistic outlook for annual growth in the G7. It is, however, highly doubtful whether these ambitious goals can be achieved within the projected time frame.

To put this in context, the IMF is projecting that the average annual growth over the 2010-2015 period will be higher in every single G7 economy compared to the rate of growth that these countries achieved in the decade of the 2000s. But, it's worth bearing in mind that the 2000s was a period where budget deficits and debt levels were far more benign than they are today.

Of course, there is always the possibility that growth which has so far been disappointing in this recovery could re-accelerate and become self-sustaining. But more sober minds know that there is no magic bullet or quick fix that one can call on to super-charge growth. When nations are faced with sizeable budget deficits and mountainous debt levels, going even deeper into debt to support the economy is a high risk strategy which will likely only add fuel to the fire and postpone the day of reckoning.

G7 Countries Face Wrenching Challenges

In the short run, the Great Recession has had a huge impact on government finances in the G7. However, in the longer run, it will be the inexorable growth in pension and health care costs associated with an ageing population that will be the major driver behind government finances.

The challenge facing the G7 will be in preventing deficits and public sector debt from growing to unsustainable levels. Inevitably, this will mean significantly raising revenues as a percentage of GDP while at the same time reducing spending levels sharply. The difficulty here is that while government's can control spending – providing there is the political will which, based on past experience, is admittedly a stretch – the intake of revenues very much depends on the state of the economy.

As fiscal policy tightens and austerity measures start to bite, the onus will be on monetary policy to keep growth going. But, with inflationary pressures mounting across the board in both developed and emerging economies, the era of ultra-low interest rates is drawing to a close. As rates start to move up this too will have a dampening effect on growth and put further pressure on government revenues.

Indeed, with recent incoming data showing that the economic recovery is starting to lose momentum in all the major economies, it is more than likely that governments will almost immediately face a short-fall in expected revenues which will stymie their plans to rein-in deficits.

The Bottom Line

As governments embark on the long and difficult road to fiscal consolidation and households struggle to pay down debt, economic activity in the G7 countries is slated to remain well below their potential levels for a considerable period of time.

29

Credit Ratings And Government Debt

— A Tale of Woe —

(July 2011)

The Great Recession of 2008-2009 that savaged the global economy and wiped out more than 50 million jobs worldwide has also blown a giant size hole in the government finances of countless countries. Public debt levels have shot up dramatically and are still rising especially in the developed economies.

In the G7, the average debt level is expected to hit 112% of GDP this year according to the IMF, up from 85% in the pre-crisis year of 2007 and in the PIGS (Portugal, Ireland, Greece and Spain) the ratio is projected to reach 105%, up from 57%. In stark contrast, the average debt level in the emerging BRIC bloc of nations (Brazil, Russia, India and China) which has better weathered the global crisis is expected to dip to 40% of GDP this year, down from 42% in 2007.

In the G7, debt ratios have increased in all the member countries and this year the United States is expected to join the ranks of the super-debtors, Japan and Italy, as its debt hits 100% of GDP. But, in terms of sheer speed, the United Kingdom grabs the title of plunging into debt at the fastest rate among this group. The country's debt-to-GDP ratio is forecast by the IMF to reach 83% this year, up from 44% in 2007 (see Table 1).

<table>
<tr><th colspan="5">TABLE 1
CREDIT RATINGS & GOVERNMENT DEBT</th></tr>
<tr><td></td><td>Rating[1]</td><td>Outlook[1]</td><td colspan="2">Debt-to-GDP Ratio[2]</td></tr>
<tr><td></td><td></td><td></td><td>(2007)</td><td>(2011)</td></tr>
<tr><td>*G-7 Economies*</td><td></td><td></td><td></td><td></td></tr>
<tr><td>Canada</td><td>AAA</td><td>Stable</td><td>67</td><td>84</td></tr>
<tr><td>France</td><td>AAA</td><td>Stable</td><td>64</td><td>88</td></tr>
<tr><td>Germany</td><td>AAA</td><td>Stable</td><td>65</td><td>80</td></tr>
<tr><td>Italy</td><td>A+</td><td>Negative</td><td>104</td><td>120</td></tr>
<tr><td>Japan</td><td>AA-</td><td>Stable</td><td>188</td><td>229</td></tr>
<tr><td>United Kingdom</td><td>AAA</td><td>Stable</td><td>44</td><td>83</td></tr>
<tr><td>United States</td><td>AAA</td><td>Negative</td><td>62</td><td>100</td></tr>
<tr><td>*BRIC Economies*</td><td></td><td></td><td></td><td></td></tr>
<tr><td>Brazil</td><td>BBB-</td><td>Stable</td><td>65</td><td>66</td></tr>
<tr><td>Russia</td><td>BBB</td><td>Stable</td><td>9</td><td>9</td></tr>
<tr><td>India</td><td>BBB-</td><td>Stable</td><td>73</td><td>68</td></tr>
<tr><td>China</td><td>AA-</td><td>Stable</td><td>20</td><td>17</td></tr>
<tr><td>*PIGS Economies*</td><td></td><td></td><td></td><td></td></tr>
<tr><td>Portugal</td><td>BBB-</td><td>Negative</td><td>63</td><td>91</td></tr>
<tr><td>Ireland</td><td>BBB+</td><td>Negative</td><td>25</td><td>114</td></tr>
<tr><td>Greece</td><td>CCC</td><td>Negative</td><td>105</td><td>152</td></tr>
<tr><td>Spain</td><td>AA</td><td>Negative</td><td>36</td><td>64</td></tr>
</table>

[1] Standard & Poor's, [2] General Government Gross Debt (%); Source: S&P, IMF

Among the PIGS, the ratio is expected to range from 64% in Spain to 152% in Greece. But, the descent into debt has been particularly sharp in Ireland. Its debt ratio has more than quadrupled, rising from 25% of GDP in 2007 to an anticipated 114% this year. It's also worth noting that Spain has a lower debt ratio than each of the G7 countries.

In the BRIC nations, debt ratios are expected to range from a low of 9% in Russia to a high of 68% in India. But what stands out is that unlike the G7 and the PIGS, debt levels in these emerging economies have not only remained under control but in the case of China and India they are in fact falling. Interestingly, government debt ratios in each of the BRIC economies are now lower than that of all the G7 countries.

Credit Rating Agencies in the Eye of the Storm

These high levels of debt are casting a dark shadow over a global recovery that is still struggling to find its legs. Worried that governments will increasingly find it difficult to meet their growing debt obligations in a low-growth environment, credit rating agencies have been particularly busy in recent weeks downgrading and issuing warnings about the sovereign debt of a number of countries.

Among the G7, Standard & Poor's, a major ratings agency, downgraded its outlook on America's debt in April for the first-time from 'stable' to 'negative' and in June it also put Italy's single A plus debt rating on negative watch (see Table 1). Although the US retained its triple A rating, the precedent breaking move by S&P means that investors can no longer assume that the US has an ironclad hold on its top-notch rating. However, should the situation worsen the dilemma for S&P is that were it to downgrade the US further this may trigger a series of events that could well be catastrophic for world financial markets.

It is worth noting that Dagong Global Credit Rating, China's main ratings agency and one that is virtually unknown in the West, has already downgraded US debt. Looking ahead, it is becoming abundantly clear that in order to attract offshore investors to hold US government debt, interest rates on US treasuries will have to rise sooner or later, perhaps appreciably.

While the debt of all the G7 countries are rated in the top A category, among the BRIC economies only China makes it into S&P's top stratum receiving a double A minus rating. Remarkably, although Japan's debt-to-GDP ratio (229%) is expected to be more than 13 times larger than China's (17%) this year, both countries have the same credit rating (AA-).

In contrast, both Brazil and India are accorded a BBB minus rating, the agency's lowest investment grade that is just a notch above junk status. Still, given their stronger growth prospects, S&P rates the long-term credit outlook for these countries as stable.

Unlike the BRICs, S&P has a negative outlook on the credit ratings for all the PIGS. And here, Greece is in a class of its own. Signaling that a default was likely, S&P downgraded Greece's long-term credit rating to junk status in June and the country now has the dubious distinction of having the lowest credit rating in the world.

Greece's on-going debt crisis is spreading and is sending the cost of borrowing zooming in the debt-laden PIGS. At the end of June, the yield on 10-year government bonds was 16.2% in Greece, 11.8% in Ireland and 11.2% in Portugal. Moreover, the fear of contagion is also pushing up interest rates in Spain. Despite having a double A credit rating and a debt ratio that is well below its peer group, the yield on 10-year Spanish government bonds has climbed well above 5%, and is at a record high since the launch of the euro.

Ultra-low interest rates have kept the cost of servicing government debt in the major advanced economies artificially low, but this state of affairs can't last forever. As government debt starts to get either downgraded or put on watch by rating agencies, borrowing costs are set to increase.

In addition, with the headline rate of inflation already rising globally, a number of central banks including the European Central Bank, the Reserve Bank of India, and the People's Bank of China have started to tighten monetary policy and are raising rates. As interest rates start to climb the cost of servicing the national debt will follow suit putting added pressure on government budgets.

The Bottom Line

The national debt in several economies is clearly on an unsustainable path and the mountain of debt which took decades to build is also going to take several years to unwind. Reducing the debt-load will require difficult policy choices and will necessitate implementing a multi-year program of fiscal restraint. But, this will not be easy.

We are now in an era where pension and health care costs are set to zoom at an unprecedented rate as the population ages. This additional burden will profoundly challenge governments and is likely to lead to considerable political uncertainty and social unrest.

Moreover, climbing out of debt will also be a major drag on growth and with unemployment rates expected to remain stubbornly high, the adjustment will be painful. The upshot is that standards of living will continue to be squeezed in the advanced economies.

The old adage *'do not spend what you do not have'* that was so royally ignored for all these years – particularly in the West – is now coming back to haunt the advanced economies with a vengeance.

30

The Faltering Recovery In G7 Economies

(August 2011)

Three years after the onset of the global financial crisis which brought several major countries virtually to their knees most are still struggling to recover from the sharp downturn. Given the significant variations in the pace of recovery, it is interesting to view the G7 economies through a macro-economic lens to see how they stack-up against each other in the current economic cycle.

If we set the benchmark at the pre-crisis year of 2007 we can easily compare the relative performance of the G7 economies over the past three years. As can be seen from the accompanying table, there have been sharp divergences in growth among this group with Canada in the lead position and Italy trailing the pack.

Canada Stands Out Among the G7

While all the G7 economies have been recovering from the global downturn, Canada is the only economy in this group that has recouped all the output it lost during the recession. Between 2007 and 2010, Canada's GDP increased at an average annual rate of 0.3

percent and the total volume of goods and services produced in the country last year was one percentage point above its 2007 peak. It's nothing to brag about for sure, but when viewed against the travails of the other major economies it is certainly an accomplishment.

Despite the return to growth, the level of output in all the other G7 economies nevertheless remains below pre-crisis levels. For example, Italy's GDP is still 5.3% below its pre-recession level. Similarly, the economies of the United Kingdom and Japan are both 3.7 percentage points smaller than they were at the start of the recession. And although Germany has had a robust recovery powered primarily by its export sector, its economy is still 0.6 percentage points smaller than it was in 2007.

TABLE 1
ECONOMIC PERFORMANCE OF THE G7 ECONOMIES

G7 Economies	Size of Economy in 2010 Relative to 2007 (in percentage points)	GDP Projections for 2011 (% change)
Canada	+1.0	2.9
Germany	-0.6	3.3
United States	-0.9	2.3
France	-1.2	2.0
Japan	-3.7	-0.6
United Kingdom	-3.7	1.3
Italy	-5.3	0.8

Source: OECD, The Economist

Although Canada has moved from recovery to expansion, its foray into the expansionary phase of the cycle could be short-lived as incoming data suggests that growth has stalled. GDP was flat in April and decreased 0.3% in May raising the risk that the economy could have contracted in the second quarter. Going forward, two key risks – the withering of growth in the US and the strong Canadian dollar – will weigh on the Canadian economy.

For Canada it will very much depend on whether growth in the United States can pick up speed and here, unfortunately, the odds of that happening do not look good. After averaging growth of 3% in 2010, the pace of recovery in the US has slowed sharply. Revised US growth figures show that the US economy averaged growth of only 0.7% in the first half of this year which is dangerously close to stall-speed.

Economic Growth Outlook Darkens in G7 Countries

Aside from the US, in the past few weeks the inflow of fresh economic data shows that the recovery in all the G7 countries has started to falter. This deceleration in growth is raising fears that the major economies could again be on the cusp of sliding into another recession.

Japan, the G7's second biggest economy, is still struggling to recover from the March earthquake and tsunami and it is also being hampered by a surging yen which is negatively affecting its export sector. With GDP shrinking by 0.3% in the second quarter the Japanese economy has now contracted for three consecutive quarters and it remains, for now at least, the only G7 economy to experience a double-dip recession.

In Germany, which has been the mainstay behind Europe's expansion, growth virtually stalled during the quarter. With GDP increasing by just 0.1% between April and June it was the German economy's weakest performance since it emerged from recession two years ago. France's recovery flat-lined in the second quarter and Italy's economy grew by 0.3% up marginally from 0.1% in the first quarter. The UK economy slowed to 0.2% in the second quarter, down from 0.5% in the previous quarter.

Based on the Economist's latest (August) poll of forecasters, growth in the G7 is expected to range from 0.8% in Italy to 3.3% in Germany this year. Japan's economy, however, is expected to shrink by 0.6% this year, the only country in the G7 to do so. Extrapolating from these projections, the German, French, and US economies are expected to join Canada and surpass their respective pre-crisis 2007 GDP levels this year.

But, the same cannot be said for the UK and Italy. Reflecting the generally weak growth outlook, it will likely be 2014 before GDP in the United Kingdom returns to its pre-crisis levels. Given its meager growth prospects, the timeline for the Italian economy is even longer and it could be 2016 or 2017 before its GDP regains its pre-recession level of output – a full decade later.

The Bottom Line

With the exception of Canada, the G7 economies still haven't fully recovered from the financial crisis that began in 2007 with the collapse of US housing and the meltdown of the sub-prime mortgage market. And, as noted above, economic growth in the major economies is now clearly running out of steam. The problem is that the governments are also running out of policy options to stimulate growth.

Indeed, after pursuing a super-loose monetary policy with rock-bottom interest rates and massive doses of quantitative easing, there is precious little that the central banks can do to jump-start growth again. Moreover, as countries press ahead with austerity measures aimed at reigning in unsustainable budget deficits and ballooning government debt levels, this will further darken the outlook for economic growth.

Unfortunately, when dealing with an atypical 'balance sheet' recession, the assumption that expansionary fiscal and monetary policy measures would underpin demand and lead to a self-sustaining recovery has so far proved elusive. Indeed, if anything, it looks like the G7 economies are beginning to seize up again. It is therefore little wonder that the economic forecasts by central banks and investment houses keep getting downgraded at a bewildering rate.

The harsh reality is that having accumulated a record amount of household and government debt over the past few decades, it could well take a generation or so to work through the debt overhang. The road ahead will not be an easy one.

31

Greece On Fast Track To Default

(October 2011)

To say that Greece's economy is in dire straits would, at this point, be a gross understatement. The country's economy is in a state of collapse and its debt is spiraling upwards.

Since being hit by the global financial crisis three years ago, GDP has shrunk by nearly 12% in volume terms and by 7% in nominal terms. At the same time its debt load has shot up, rising from €262 billion in 2008 to €366 billion this year. Relative to the size of its economy, Greece's debt-to-GDP ratio has climbed from 110% to 166% and is by far the highest in the Eurozone (see table 1).

Furthermore, the country is uncompetitive inside the Eurozone, has a large current account deficit (-8% of GDP), can't devalue the currency to restart its economy, and interest rates are at astronomical levels (the 10-year bond is trading at a yield of around 23% and the 2-year yield is at 50%). Being completely shut out from raising funds in international capital markets the government's only recourse has been to go cap-in-hand and seek bailouts from the international community including the International Monetary Fund, the

European Central Bank and the European Union, the so-called 'troika'.

TABLE 1
THE ROAD TO DEFAULT
Greece's Economy is Collapsing while its Debt is Spiraling Upwards

	2008	2011	% Change
Real GDP	€184.0bn	€163.3bn	-11.3%
Nominal GDP	€236.9bn	€220.9bn	-6.8%
Government Debt	€262.3bn	€365.6bn	+39.4%
Debt-to-GDP Ratio (%)	110.7	165.6	+49.6%

Source: IMF

Given Greece's rapidly deteriorating economic fundamentals with output continuing to shrink and the unemployment rate climbing inexorably, no amount of bailouts will put Greece back on the road to economic health. With the economy caught in an ever-tightening straitjacket, it will take a miracle for the country to regain its economic footing. And, here's the rub: even if the government manages to balance its budget by 2013 and does not need fresh capital to fund government spending, the country still has to repay huge debts.

New Austerity Measures

The 'troika' of lenders is now demanding that in order to secure the next bailout tranche of €8 billion the Greek government must enact further measures to cut its borrowing requirements. To avoid defaulting on its current €366 billion debt the authorities are pushing through an additional set of austerity measures that includes slashing pensions and public sector wages. A new property tax, to be paid through electricity bills to make it easier and faster for the state to collect, has also been instituted.

This latest bailout tranche, the sixth installment of a €110 billion bailout package that was agreed to in May 2010, is expected to be approved, albeit with delays, and will give Greece the wiggle room it needs to continue paying its bills including civil service salaries and to keep it afloat. But, it's only a short-term palliative.

The only way for Greece to solve its debt problem is through economic growth. Raising taxes and cutting spending when the economy is mired in a recession will only deepen the slump and make it impossible for it to work off its debt.

What is incongruous is that public sector salaries, instead of being afforded through tax revenues which is generally the case, are being funded through borrowing. Moreover, what is sure to stick in the craw of the countries that are being urged by their governments to pony up even more funds is that under the Greek Constitution public sector workers are essentially guaranteed jobs for life.

Facing Harsh Realities

The choices facing the Greek government are stark. Either it continues to heap even more austerity measures on its citizens in order to get additional bailout funds from the 'troika' or it pulls the plug – defaults – and leaves the euro. But, borrowing ever greater sums of money to shore up an economy that is spiraling downwards simply postpones the day of reckoning. If it chooses to maintain the status quo there is little that Greece itself can do to escape from its crushing debt burden. It can sell some of its assets to pay down part of its debt, but given its enormous size that will hardly make a dent.

Being part of the Eurozone, Greece cannot pursue an independent monetary policy and inflate away its debt nor does it have the escape route of devaluing the currency to try and jump start its moribund export sector. As long as it shares the common currency, the only way for Greece to regain its competitiveness is through internal devaluation which entails lowering its cost structure and cutting wages. But, the failure of repeated rounds of spending cuts and tax hikes to put the country's finances on a more even keel is now starting to grate on the public.

Not surprisingly, the Greek population is increasingly getting more agitated and as the fiscal screws keep getting tightened the public's patience may be reaching a breaking point. So, unless household demand suddenly revives and the private sector kicks in, the odds of which are, let's face it, flat-out zero, the only way for Greece to restart its economy and regain its competitiveness is for it to leave the euro – the costs of which would be formidable both to Greece and the Eurozone.

Nevertheless, the Greeks may decide that abandoning the euro and reintroducing their own currency – the Drachma – will allow them to regain their economic sovereignty and march to their own drumbeat. Either way, the path to regaining its economic health is going to be long and painful.

Greece Enters the Dark Zone

Greece's economy is now trapped in a downward spiral from which it cannot extricate itself. The austerity measures demanded by the 'troika' before releasing additional funds will push the Greek economy even further into an ever-deepening hole. As deflationary forces tighten their grip on the economy and demand continues to plunge and unemployment surges, it will be all but impossible for Athens to meet its debt repayments.

The only way for Greece to avoid defaulting on its debts would be for the 'troika' to keep on lending it money. But, this is a non-starter. Greece cannot be subsidized forever. It is therefore just a matter of time before Greece defaults on its debts which in turn could set-off a domino effect whereby other heavily indebted Eurozone countries including Ireland, Portugal, Spain and perhaps even Italy start to unshackle themselves from the euro.

Depending on how things play out on the political front a Greek default could be the trigger that leads to a messy break-up of the Eurozone.

In a country that is already mired in a deep recession with mass unemployment – one in six Greeks is currently unemployed – layering on even more tax increases and spending cuts will lead to an escalation of social unrest. Indeed, Greece's major labor unions have already called for another general strike for October 19. So far, the demonstrations and protests have been contained but as the economy continues to sink and more and more people are thrown into poverty, the situation in Greece could become explosive. As the mountain of debt continues to grow, the latest tax hikes and spending cuts could be the final nail that leads to the total collapse of Greece's economy.

32

The Looming Break-Up Of The Eurozone

(December 2011)

The debt crisis that has been plaguing Europe for several months has been spreading and intensifying. It is now entering a dangerous phase where serious doubts are being raised among investors about the very survival of the euro. At the December 8-9 Brussels summit – the latest in a string of high-level meetings – all the 17 Eurozone countries agreed to a new 'fiscal compact' that is designed to allow Brussels to enforce stricter fiscal discipline and impose automatic penalties on recalcitrant members. However, the initial optimism of the summit's outcome lasted only a few short hours and has now all but vanished.

While the goal of fiscal union to go along with monetary union is the correct solution to fixing what ails the Eurozone, it will require changes to the EU treaty which can't be done overnight. Moreover, a successful outcome is far from guaranteed. Indeed, it is more than likely that without an electorate mandate nations will not acquiesce quietly to having to hand over much of their economic sovereignty to Brussels. Inevitably, many contentious political battles lie ahead.

As there are no quick-fixes to the crisis that is raging, financial markets remain on edge concerned that politicians have all but run out of policy options. Realistically, there is unlikely to be any sort of deal that can be implemented immediately to backstop the Eurozone from falling apart. To stave off these countries from defaulting and collapsing, the Eurozone governments must implement emergency measures now.

Debt Levels and GDP

In the short-term the only way out of this morass is for the European Central Bank to act as a lender of last resort by buying the sovereign debt of the heavily indebted Eurozone countries in unlimited quantities. But, the Germans and the ECB have categorically ruled this out arguing that this would be in violation of the EU Treaty.

TABLE 1
EUROZONE NATIONS IN THE 'DARK ZONE'

	Budget Deficit[1]	Government Debt[1]	GDP Growth[2]	Unemployment[3]
Greece	-7.0%	181%	-7.5%	18.3%
Portugal	-4.5%	122%	-4.1%	12.9%
Ireland	-8.7%	119%	-0.2%	14.3%
Spain	-4.4%	77%	-0.7%	22.8%
Italy	-1.6%	128%	-1.0%	8.5%

[1] as a percent of GDP, OECD projections for 2012; [2] The Economist Poll for 2012 (December 10, 2011), [3] October 2011, except Greece August 2011

Source: OECD, The Economist

However, even if the ECB were to step in and buy the debt of individual countries it would only be a stop-gap measure. The fundamental problem is that the level of debt, particularly among the peripheral countries of Greece, Portugal, and Ireland, is growing at a much faster rate than that of their economies.

These economies, which are already on life-support and are receiving bailouts from the IMF, the European Central Bank, and the European Union, are in a class by themselves. They are saddled with large budget deficits, mountainous debt levels, and Great Depression levels of unemployment. But what is really the crippling factor is that GDP continues to fall (see table).

Indeed, Greece's GDP has declined for four consecutive years and the economy is now over 17 percentage points smaller than what it was before the onset of the Great Recession in 2007. Ireland's economy is 11 percentage points smaller and Portugal's is over 7% smaller.

To add to Europe's woes, Spain and Italy have now also joined Greece, Portugal and Ireland on the critical list and, with their economies heading back into recession (see table), it is looking increasingly likely that they too may need a bailout to stay afloat. Indeed, in recent weeks Italian and Spanish 10-year government bond yields have been rising sharply in volatile trading and are at near record highs since the introduction of the euro in 1999.

The Curse of Being Uncompetitive

Without economic growth, governments are already finding it extremely difficult to service their debts and are caught in a vicious spiral where interest rates are surging and debt levels continue to grow. Moreover, these 'critical list' economies are uncompetitive.

Over the ten year period from the introduction of the euro in 1999 until the end of 2009 unit labour costs in Portugal, Ireland, Greece, Spain and Italy have risen on average by over 30%. In sharp contrast, Germany has kept a tight lid on its wages and unit labour costs have increased by less than 4% over the same period. It's little wonder that Germany remains super competitive.

The inherent problem is that being part of the Eurozone these economies cannot pursue an independent monetary policy and inflate away their debt nor can they devalue the currency to try and regain their competitiveness and jump-start their exports. As long as they share the common currency, the only way for the PIIGS to be competitive is through internal devaluation which entails lowering their cost structure and cutting wages. But, the failure of repeated rounds of spending cuts and tax hikes to put their finances on a more even keel is leading to social unrest.

A Painful Choice

Unable to formulate their own independent monetary and exchange rate policies to lift their respective economies from a severe recession, the PIIGS face a wrenching dilemma. Either they remain within the Eurozone and endure economic stagnation, high unemployment, and sharply falling standards of living on the back of endless austerity measures, or they break away from the Eurozone and take control of their economies.

Greece, Portugal, and Ireland are the three most likely Eurozone countries to throw in the towel and leave the euro. Should this exodus occur it would unleash deflationary forces around the globe and the ensuing credit crunch could lead to a run on the banks akin to what happened to Northern Rock in 2008 when the British government was forced to step in and nationalize the bank.

The impact of the meltdown in financial markets would immediately be felt on the real economy. Europe's economy would undergo a severe recession as demand collapsed which in turn would send the unemployment rate soaring. As Europe's economy implodes it would send shock waves through the global economy. The US economy would feel the full force primarily through the financial system and China's economy would quickly suffer a hard landing as demand for its exports collapses.

A New Euro

Perhaps the euro will survive, but it will be in an altered form. While some of the peripheral economies in the PIIGS might throw in the towel and exit the euro, the key Eurozone countries of Germany and France together with the likes of the Netherlands, Finland, and Austria could form the core of a revamped euro. Only time will tell.

Nevertheless, history may well show that having created the European Union with a single market of 480 million consumers, and the free movement of goods, services, capital, and people within the member countries, the establishment of a monetary union and the creation of the euro without fiscal and political union was doomed to fail.

PART 4

The Year 2012

33

E7 Growth Performance Trumps G7

(January 2012)

It is now almost three years since the Great Recession ended and profound changes are underway in the world economy.

The global economic axis which had been shifting fundamentally away from the advanced economies of Europe and North America to the world's emerging economies has accelerated sharply over the past four years. Moreover, living standards are rebalancing across the world, rising in the emerging countries but falling in the advanced countries.

The global economy has been severely buffeted in the past few years as it lurches from one crisis to yet another. The bursting of the US housing bubble, the meltdown of the sub-prime mortgage market, the freezing of credit markets, the collapse of Lehman Brothers, the sovereign debt crisis, credit rating downgrades, and the very survivability of the Eurozone, have all contributed to the unprecedented battering that is plaguing the global economy.

It's little wonder that the fallout from these crises has had a profound effect on the structure of the world economy. Interestingly, a comparison between the major advanced economies of the G7 and the seven largest emerging economies (the E7) reveals some startling differences. Collectively, the E7 bloc which includes China, India, Indonesia, Brazil, Russia, Turkey and Mexico now accounts for close to 31% of world GDP, up from 19% twenty years ago. During this same time period, the G7 has seen its share of world output fall from 51% to 38%.

TABLE 1
VITAL SIGNS: G7 vs E7
(Nominal GDP* 2011)

G-7 Countries	$Bn.	E-7 Countries	$Bn.
United States	15,065	China	11,316
Japan	4,396	India	4,470
Germany	3,089	Russia	2,376
United Kingdom	2,254	Brazil	2,309
France	2,217	Mexico	1,659
Italy	1,829	Indonesia	1,123
Canada	1,391	Turkey	1,055
% of World Total	38.4%		30.8%

* Purchasing power parity, billions of USD; Source: IMF

The impact of the global recession on the G7 and E7 economies has been quite varied. In a nutshell, while the recession and the ongoing economic malaise have knocked the wind out of the G7 economies, the impact on most of the E7 countries has been relatively muted. Five of the G7 economies – Britain, France, Italy, Japan, and the United States – all suffered back-to-back declines in GDP both in 2008 and 2009. Canada and Germany, however, posted declines in GDP on a calendar year basis only in 2009.

In contrast, four members of the E7 group – Brazil, Mexico, Russia, and Turkey – experienced declines in economic activity only in 2009 with the fall in GDP ranging from a low of -0.6% in Brazil to a high of -7.8% in Russia. Moreover, the economies of China, India, and Indonesia rode out the financial storm and sailed through the global recession without posting a single negative year of growth.

Since climbing out of the Great Recession, the recovery has been weak across the board for all the G7 economies and there are growing fears that another economic downturn may be unavoidable. For example, for the G7 group as a whole, growth in GDP averaged 2.7% in 2010 but weakened to 1.3% in 2011 and is expected to slip even further and average just 0.6% this year.

In contrast, while a slowdown is also anticipated in all the major emerging economies because of the global inter-linkages, there is no talk of recession. Economic growth in the E7 averaged 7.5% in 2010, 6.0% in 2011 and is projected to slip to 5.2% this year.

It is these divergent trends in growth that have significantly altered the global economic landscape. To put things in perspective, over the four year period from the end of 2007 through to 2011, only four of the G7 economies have regained their pre-recession levels of output.

Canada has been the best performer in this group but despite that it is still only 3.1% larger than it was in 2007. The size of Germany's economy, the second best performer, is 1.8% larger while the United States and French economies have just managed to move ahead of where they were in 2007.

Three of the G7 economies – the United Kingdom, Japan, and Italy – have failed to recover the output lost from the 2008-09 recession and find themselves essentially stuck in what amounts to a long drawn-out economic slump. The UK economy is 2.6% smaller than it was in the pre-recession peak year of 2007, Japan's is 4.2% smaller, and Italy's is 4.7% smaller (see Table 2).

In contrast to the G7 countries, the production of goods and services is bigger today in all the E7 economies than it was in 2007. China's economy is 44.6% larger than it was before the crisis and despite a slowing down of growth its GDP is likely to expand by another 8.2% this year. Similarly, India's economy is 34.6% larger, Indonesia's is 25.2% and Brazil's is 16.5% bigger. Even Mexico's economy, which is 3.9% larger and, therefore is the E7's worst performer, has outperformed every single member of the G7.

TABLE 2 THE WORLD ECONOMY RECALIBRATES (Cumulative change in Real GDP between 2007 and 2011, in percent)			
G7 Countries	% chg	E7 Countries	% chg
Canada	+3.1	China	+44.6
Germany	+1.8	India	+34.6
United States	+0.6	Indonesia	+25.2
France	+0.1	Brazil	+16.5
United Kingdom	-2.6	Turkey	+11.2
Japan	-4.2	Russia	+5.2
Italy	-4.7	Mexico	+3.9

Source: IMF

The major advanced economies now face years of struggle and none of them are likely to see a return to pre-crisis rates of growth for the next few years. Indeed, several of the G7 economies including Britain, France, Germany, and Italy could be heading back into recession as the recovery is increasingly showing signs of coming unstuck.

Unemployment is rising again in Europe, retail sales are falling, and although inflation has started to edge down it still remains above central bank targets. Moreover, the need to reduce budget deficits and reign in unsustainable debt-to-GDP ratios – which are at alarmingly high levels in all the G7 economies – risks further entrenching the recessionary conditions in which these economies find themselves stuck.

With the outlook for growth diverging sharply, the G7 countries are split into two camps – the United States and Canada are expected to grow at around 2% in 2012 and Japan's economy is also likely to see its output rise by a similar amount as the country rebuilds from last year's devastating tsunami and earthquake.

On the other hand, the outlook for European economies is darkening. With the debt crisis in the Eurozone countries continuing to swirl and showing no sign of easing, the IMF in its latest forecast expects the region's GDP to contract by 0.5% this year. Italy, the regions third largest economy is projected to decline by 2.2%, by far the worst performer of any G7 economy.

It is now abundantly clear that, more than two years after the end of the Great Recession, a sustained recovery remains stubbornly elusive for the major advanced economies. Despite massive amounts of monetary and fiscal stimulus, the rate of growth in all of the major advanced economies has been sharply below their respective long-term averages.

Moreover, constrained by large debts and deficits, not a single G7 country is expected to achieve growth rates above, or even at, it's long-term average for several more years.

In contrast, since 2007, growth in the economies of the E7, despite the ongoing global turbulence, has not deviated much from their long-term averages. By 2020 this bloc, given the current trends, will surpass the G7 and account for a greater share of world output. This, in turn, will lead to a shift in the current geo-political power structure. Whether this will be muted or more pronounced remains to be seen.

34

Canada – The Mouse That Roared

(February 2012)

The impact of the Great Recession and the subsequent recovery on the economies of Canada and the US is a study in contrasts. Unlike the last big recession of 1981-82 which hit Canada's economy much harder than that of the US, this time, instead of being dragged down by its biggest trading partner, Canada's economy has significantly outperformed its southern neighbour on both the output and employment front. The country's solid banking system, the burgeoning demand for its natural resources by emerging economies, and a housing sector that virtually sailed unscathed through the downturn were all factors that helped to mitigate the impact of the global recession on Canada.

Canada Outperforms the United States

Between the fourth quarter of 2007 and the second quarter of 2009, GDP contracted by 3.7% in Canada whereas in the US output fell by 5.1%. Moreover, the recovery has also been slower in the US. From the middle of 2009 to the third quarter of 2011, the US economy grew by only 5.5% while in Canada during this same period GDP increased by 6.7%.

In addition, Canada's economy regained its per-recession peak level of output in the third quarter of 2010, a full year ahead of the US economy, and is now 2.8% larger than it was before the recession began.

In contrast, the US economy has only now just managed to regain its pre-recession peak. Canada's job market has also outperformed that of the United States by leaps and bounds. During the recession, employment in Canada fell by 2.5% but by 6.4% in the US. The recovery also has been much stronger in Canada with job gains of 3.7% compared to the US's 2.4% (see table).

TABLE 1
DIVERGING FORTUNES

	Recession[1/] Job Loss			Recovery[2/] Job Gains	
	'000s	%		'000s	%
Canada	431	-2.5%	Canada	613	+3.7%
USA	8,779	-6.4%	USA	3,165	+2.4%

[1/]Canada: employment peak: October 2008; trough: July 2009 (9 mos.); USA: employment peak: January 2008; trough: February 2010 (25 mos.)
[2/]Recovery: from respective trough to January 2012

Source: Statistics Canada, US Bureau of Labor Statistics,

On the back of a stronger pick up in output, it took 18 months for the Canadian economy to recoup all the jobs that were lost during the downturn. Since bottoming out, the level of employment in Canada is now close to 17.4 million, a full percentage point above its October 2008 peak. Canada's unemployment rate has also steadily dropped but it still remains above the level it was before the recession began.

Employment in the US peaked in January 2008, a month after the official start of the recession, and by the time it bottomed out 25 months later in February 2010 the economy had shed close to 8.8 million jobs (see table). With the tepid US recovery, the economy has generated only about 3.2 million jobs over the past two years and has recovered just over one-third of the jobs lost in the recession.

While employment in Canada is at a historic high, the US is still 5.6 million jobs short of its pre-recession peak level and even if it maintains the average monthly gain of 138,000 since March 2010, it won't be before the middle of 2015 before the US regains its previous peak – more than eight years after the onset of the recession.

Canada's Economy Starting to Lose Momentum

Canada's roar has been somewhat muted over the past few months as the economy has hit a soft patch. Real GDP was flat in October and declined 0.1% in November which points to a sharp slowdown in economic activity. It is likely that when the fourth quarter GDP numbers are released by Statistics Canada in early March it will likely show that the pace of growth fell to around 1.5% from 3.5% in the previous quarter. On the employment front, the economy has lost 34,000 jobs since September 2011 and the unemployment rate has now ticked up for four consecutive months, rising from 7.2% in September to 7.6% in January 2012.

In contrast to Canada's slowing economy, the pace of growth in the US has been picking up. The latest figures show that the American economy expanded at an annualized 2.8% in the fourth quarter of last year, a full percentage point higher than in the previous quarter. Similarly, job creation has also picked up speed.

In just the last four months alone, from September through to January this year, the US economy has created 715,000 jobs, which represents a quarter of all the jobs gained since employment started to recover in March 2010. Furthermore, the unemployment rate has ticked down for four straight months falling from 9% in September to 8.3% in January of this year.

The improving US economy couldn't have come at a better time for Canada. It should be a boon for the country's exports and, hopefully, this will rekindle its growth trajectory.

35

Central Banks & Quantitative Easing

— *Time To Change Course* —

(May 2012)

When the global financial crisis erupted following the collapse of Lehman Brothers in 2008, the world's major central banks moved quickly to bolster the world economy and prevent it from falling into a credit-led depression. By sharply lowering interest rates in a series of cuts to record low levels they succeeded in preventing another 1930s style Great Depression from taking hold. But, despite the persistence of ultra-low interest rates over the past several years, growth in the advanced countries has failed to get onto a firmer footing.

Unable to push rates below the zero bound, the world's major central banks have been turning on the electronic spigots and flooding the global financial system with liquidity through a policy of quantitative easing. In 2007, the balance sheet assets of the major central banks including the Federal Reserve, the European Central Bank, the Bank of England, and the Bank of Japan amounted to about $4 trillion. Four years later, that figure has skyrocketed to over $9 trillion.

Since the financial crisis, the Federal Reserve has more than tripled the size of its balance sheet while the other major central banks have doubled theirs. As a percent of GDP, the balance sheets of the major central banks now range from 19% in the case of the Federal Reserve (up from 6% in 2007) to 32% for the European Central Bank, up from 17% (see table 1).

By creating new money to buy up government debt and other financial securities from commercial banks and other financial enterprises, the goal of QE was to boost economic growth, increase inflation, and lower the unemployment rate. However, when one examines the impact of all the QE that the major central banks have engaged in over the past three years, it is apparent that the policy has misfired when it comes to boosting the economy.

After an initial spurt, economic growth has started to slow sharply in the advanced countries, job creation has been weak, and the unemployment rate is still well above pre-recession levels.

TABLE 1
CENTRAL BANK BALANCE SHEET ASSETS

	Percent of GDP		USD (Trillions)	
Central Bank	2007	2011	2007	2011
Federal Reserve	6%	19%	0.8	2.9
European Central Bank	17%	32%	2.1	4.3
Bank of England	7%	22%	0.2	0.5
Bank of Japan	22%	31%	1.0	1.8

Source: IMF, WSJ

The bottom line is that the repeated rounds of bond buying have failed to put the economy onto a self-sustaining growth path. In 2010 – the first full year of growth for all the G7 countries since the Great Recession – GDP ranged from a low of 1.4% in France to a high of

4.4% in Japan. However, after that initial spurt, economic growth not only has been slumping in the advanced economies but several European economies including the United Kingdom, Italy, the Netherlands and Spain have now all fallen back into recession.

This year, the consensus forecast is for growth in the G7 to average 0.8%, down sharply from 2.8% in 2010, and range from a low of – 1.9% in Italy to a high of only 2.1% in the US and Canada.

The policy of pumping money into the economy through quantitative easing is a massive experiment that has clearly failed to shore up economic growth. While financial markets have benefited from QE – stock prices are up from their March 2009 lows, interest rates have declined, and commodity prices have been boosted – the impact on the real economy has been much more muted.

The problem is that QE is a very indirect way of trying to stimulate the economy. While QE has helped to recapitalize banks, which continue to focus on repairing and purging their balance sheets from toxic assets, it has done little to help the flow of credit to the real economy.

Despite a super-accommodative monetary policy that the central banks have pursued over the past several years, credit conditions still remain tight particularly for small and mid-sized companies. Although the whole gamut of monetary policy needs a re-think, central bankers appear determined to stay the course.

Indeed, the Federal Reserve has indicated that the federal funds rate will remain at near zero levels until well into 2014. And, there is talk that they may be preparing for another round of QE in the (vain) hope that flooding the system with even more money will eventually lead to an upsurge in growth.

Trying to force feed economic growth by saturating the system with trillions of dollars when households and governments are deleveraging has proved to be an ineffective way to spur the recovery. A better approach would be for the central banks to intervene directly into the economy by lending money to small and mid-sized companies who in the main are the lead generators of employment. One key consequence of credit being in short supply is that over time it lowers the economy's potential rate of growth.

While buying government bonds directly has led to lower interest rates and benefited the banking sector and lowered the debt servicing costs for governments, it has also driven down the returns for millions of retirees.

After three years of record low interest rates, the central banks have indicated that they intend to keep rates low for years to come. Unfortunately, this means that for savers and pensioners the struggle to stay afloat will continue as record low interest rates translate into lower incomes.

For large swathes of baby boomers, a comfortable retirement is now out of the question.

36

Mounting Debt Levels To Overwhelm G7 Economies

(September 2012)

That the global financial crisis and the ensuing Great Recession of 2008-2009 have wreaked havoc on the public finances of the major advanced economies is not news. What is news though is that the debt-to-GDP ratios in all the G7 economies are projected to be considerably higher in 2017 – a full ten years after the onset of the crisis.

An analysis of the IMF's latest medium-term projections shows that over the 2007-2017 period the level of debt is projected to increase at a faster pace than growth in output in each of the major economies further pushing up the debt-to-GDP ratio. In 2007, debt ratios ranged from a low of 44% in the UK to a high of 183% in Japan. But, with the level of debt projected to grow faster than their respective economies, the debt-to-GDP ratio is slated to be higher in all G7 nations in 2017 compared to 2007. By 2017, the ratio is expected to range from a low of 71% in Germany to a high of 257% in Japan (see table).

In nominal terms the total value of goods and services produced is projected to be larger in all the G7 countries in 2017 compared to a decade earlier, but growth rates are expected to vary widely. For instance, the IMF projects that by 2017 Japan's economy will only be 1% larger than it was in 2007 whereas at the other end of the spectrum, Canada's economy will be 46% bigger, the United Kingdom's 42% and the United States at 40%.

The major European economies are also projected to grow but at a considerably slower pace. For example, France's economy is estimated to be 31% larger followed by Germany at 24% and Italy at 12%. Since 2007, national debt levels have been exploding in all the major advanced economies. By 2017, the level of government debt is expected to more than double in the US, rising from $9.4 trillion dollars in 2007 to $22.3 trillion.

		\multicolumn{2}{c	}{TABLE 1 DROWNING IN DEBT}		
		Trillions		% of GDP	
G-7 Countries	Currency	2007	2017	2007	2017
Canada	CAD	1.02	1.64	67	74
France	EURO	1.21	2.09	64	85
Germany	EURO	1.58	2.14	65	71
Italy	EURO	1.60	2.07	103	119
Japan	YEN	93.9	132.9	183	257
United Kingdom	GBP	0.62	1.73	44	87
United States	USD	9.42	22.26	67	113

Source: IMF

The debt outlook for the UK is even more dramatic. That country's national debt is projected to hit £1.7 trillion in 2017, a jump of £1.1 trillion from its 2007 level. Canada's debt level is expected to top $1.6 trillion in 2017, up from about $1 trillion in 2007. In France, Germany and Italy, debt levels are projected to top €2 trillion in each country and Japan is expected to tack on another ¥39 trillion in debt.

With public finances that are already at abysmal levels, the projected deterioration in the debt levels of all the major advanced economies will continue to challenge the ability of governments to service these debt levels. The best way out of the debt trap is via economic growth but, alas, despite letting loose record amounts of monetary and fiscal stimulus, growth has failed to gain traction. Indeed, the latest data shows that GDP is heading down not only in the G7 economies but it is also decelerating sharply globally.

The Bottom Line

The winding down these mountainous debt levels will take a long time and will act as a brake on economic growth. It looks like this decade will be the decade of the 'Great Stagnation' in the G7 economies.

37

Current Policy Path Dooms The Eurozone

(November 2012)

The Euro-zone is at a breaking point and, given its current economic policy path, it is unlikely to survive intact for long. Several economies including Greece, Ireland, Portugal, and Spain are mired in a deep recession and it is not at all clear when or how they will return to growth.

As one can see from the table below the statistics make for grim reading. These economies are shrinking dramatically. Demand is spiraling down, unemployment rates are shooting up and debt levels are in the red zone where alarm bells are ringing the loudest. Greece in particular is stuck in a severe economic depression.

Since 2008, domestic demand has fallen by 25.8% and the jobless rate has tripled to 25.4%. By way of comparison, during the Great Depression domestic demand in the United States fell by 24% between 1929 and 1933 and the unemployment rate peaked at 24.8% in 1933.

The economic and social cost of the 2008/09 global recession and the subsequent imposition of austerity measures to rein in budget deficits have been profound. Buried in debt and unable to borrow to meet their financing needs from international credit markets, Greece, Ireland and Portugal have each been bailed out by the Troika (European Union, European Central Bank, and the International Monetary Fund).

But, in return for the bailouts, the Troika has demanded a host of structural changes ranging from labour market reforms to deep cuts in government spending. These austerity measures have included across-the-board cuts to public sector wages and pensions, a reduction in welfare benefits and higher taxes.

	Budget Deficit	Debt-to-GDP Ratio	Domestic Demand[1]	Unemployment Rate[2]
	(2012)	(2012)	(2008-12)	Total
Greece	-7.5%	170.7%	-25.8%	25.4%
Spain	-7.0%	90.7%	-13.5%	25.8%
Portugal	-5.0%	119.1%	-13.9%	15.9%
Ireland	-8.3%	117.7%	-21.7%	15.1%

**TABLE 1
SINKING INTO A 'GREAT DEPRESSION'**

[1]Cumulative change, [2]September 2012 except Greece July 2012.
Sources: IMF, OECD, Eurostat

The problem is that these measures have not only failed to shore-up confidence in financial markets but they have ended up destroying any potential for growth. As households have been forced to drastically cut back their spending, businesses have followed suit and slashed employment and investment.

This, of course, is driving up the unemployment rate which further shrinks the tax base, pushes up the deficit and further adds to the level of debt. It's little wonder then that the bailout recipients have been unable to meet their debt repayment targets.

Despite all the efforts to stabilize the financial markets and revive the economy, what the austerity measures have achieved so far is that they have succeeded in pushing these countries into an economic depression. Moreover, the situation is getting worse. As the level of debt continues to climb, the imposition of further austerity measures will only compound the problem and deepen the pace of the contraction.

These economies are caught in a vicious circle with no escape route from the strait-jacket that they find themselves in. The fundamental problem is that they are basically uncompetitive and cannot grow their way out of the debt crisis. Moreover, being members of the currency union, they can't devalue the currency in order to restore competitiveness and boost export demand.

These countries are unraveling, not just economically but politically and socially as well. Confidence among the electorate has been shattered and they face a stark choice: either they stay in the euro, accept the bailout conditions and cede their economic sovereignty to Brussels or they abandon the euro, re-establish their own currency, take control of their own economic destiny and face the consequences of being shut out from international markets for years. Either way it is going to be very painful.

A Call for a Policy Re-Think.

The top priority for the governments should be to stabilize the economy and provide a more predictable economic environment so that growth can resume. But this requires a new approach to tackling the crisis. This can be achieved by a combination of measures including, for example, imposing a moratorium on any additional austerity measures for a minimum of, say, three to five years and instigating a multi-year freeze on public sector wages and pensions.

To avoid outright defaults a large scale restructuring of debt is called for. This could include a lengthening of the term structures, delaying interest rate payments and some level of debt forgiveness. On the trade front the imposition of a special tax on imports would achieve two goals. It would help boost domestic demand by redirecting spending towards the domestic economy and at the same time help to correct the balance of payments deficits that these countries are running.

By sticking to the current austerity measures, it is difficult to see how Greece, Portugal and Spain can escape from the current Euro straightjacket. The combination of tight fiscal policy, volatility on the interest rate front, and an overvalued euro from the perspective of the southern euro-zone members has resulted in pushing Greece et al into an economic depression.

Saddled with huge debt loads that continue to climb and with GDP shrinking, these economies are unable to generate enough tax revenues to service their debt. To all intents and purposes they are well past the point of debt saturation and are in effect bankrupt.

Given these facts, it is difficult to understand why Europe's policy makers are persisting with such a doctrinaire-based policy that has clearly failed to put Europe back onto a path of steady growth. As Winston Churchill once famously noted: 'However beautiful the strategy, you should occasionally look at the results'.

With a crippling economic depression that currently is spreading across the Eurzone it is high time for Europe's policy makers to change course. The very future of the Euro-zone hangs in the balance.

PART 5

Articles Published by

The Globe and Mail

Canada's National Newspaper

I

G7 REVENUES AND SPENDING: MIND THE GAP

(Published June 23, 2011)

The financial crisis and the ensuing global recession have dealt a crippling blow to public sector finances in the G7 and from which all member countries have yet to recover. As GDP contracted and the unemployment rate surged, government revenues fell and spending jumped in all the major economies but with wide variations.

As a result of these sharp divergences in government revenues and expenditures, the fiscal deficit has risen in all the major economies. Before the onset of the global recession, only Canada and Germany were running surpluses but by 2009, all the G7 economies were in the red.

According to the International Monetary Fund's latest medium-term projections, over the 2010-2015 period, government revenues are projected to increase at a faster pace than the overall growth in the economy in all of the G7 countries. On the other hand, government spending is projected to grow at a considerably slower pace in all cases. This growing gap is expected to sharply reduce the budget deficits of the major economies. Still, by 2015, only the budgets in Canada and Germany are expected to be in balance. In the rest of the group the deficit, as a percentage of GDP, is projected to be 7.4 per cent in Japan, 5.5 per cent in the United States, 3.1 per cent in Italy, 2.3 per cent in the United Kingdom, and 2.2 per cent in France.

The question, of course, is how realistic are these projections?

The GDP forecasts undoubtedly are the wild card and here the IMF clearly has an optimistic outlook for annual growth in the G7. To put this in context, the IMF is projecting that the average annual growth over the 2010-2015 period will be higher in every single G7 economy compared to the rate of growth that these countries achieved in the 2000s. But, it's worth bearing in mind that the 2000s was a period where budget deficits and debt levels were far more benign than they are today.

The challenge facing the G7 will be in preventing deficits and public sector debt from growing to unsustainable levels. Inevitably, this will mean significantly raising revenues as a percentage of GDP while at the same time reducing spending levels sharply. The difficulty here is that while governments can control spending -- providing there is the political will which, based on past experience, is admittedly a stretch -- the intake of revenues very much depends on the state of the economy.

As fiscal policy tightens and austerity measures start to bite, the onus will be on monetary policy to keep growth going. But, with inflationary pressures mounting across the board in both developed and emerging economies, the era of ultra-low interest rates is drawing to a close. As rates start to move up this too will have a dampening effect on growth and put further pressure on government revenues.

The bottom line is that as governments embark on the long and difficult road to fiscal consolidation and households struggle to pay down debt, economic activity in the G7 countries is slated to remain well below their potential levels for a considerable period of time.

II

The Price of Debt: Higher Borrowing Costs

(Published July 19, 2011)

Public debt levels have shot up dramatically since the Great Recession and are still rising, especially in the developed economies. Debt ratios have increased in all the G7 countries, and this year the United States is expected to join the ranks of the super-debtors -- Japan and Italy -- as its debt hits 100 per cent of GDP. But, in terms of sheer speed, the United Kingdom grabs the title of plunging into debt at the fastest rate among this group.

The country's debt-to-GDP ratio is forecast to reach 83 per cent this year, up from 44 per cent in 2007. In Canada's case the debt-ratio is projected to climb from 67 per cent to 84 per cent. In the BRIC nations, debt ratios are expected to range from 9 per cent in Russia to 68 per cent in India and, interestingly, will be lower than that of all the G7 countries.

Credit rating agencies, worried that governments will increasingly find it difficult to meet growing debt obligations in a low-growth environment, have been particularly busy in recent weeks.

Among the G7, Standard & Poor's downgraded its outlook on America's debt in April for the first-time from 'stable' to 'negative' and in June it put Italy's A+ debt rating on negative watch.

It is worth noting that Dagong Global Credit Rating, China's main ratings agency, has already downgraded U.S. debt. While the debt of all the G7 countries are rated in the top A category, among the BRIC economies only China makes it into S&P's top stratum, receiving an AA– rating. Remarkably, although Japan's debt-to-GDP ratio (229 per cent) is expected to be more than 13 times larger than China's (17 per cent) this year, both countries have the same credit rating (AA–).

When we look at the PIGS (Portugal, Ireland, Greece and Spain), S&P signaled that a default was likely by downgrading Greece's credit rating to junk status in June and the country now has the lowest credit rating in the world. Moreover, Greece's on-going debt crisis is spreading and the cost of borrowing is zooming in the debt-laden PIGS. In Spain, despite having an AA rating and a debt ratio that is well below its peer group, the yield on 10-year Spanish government bonds has climbed well above 5 per cent -- a record high since the launch of the euro.

Ultra-low interest rates have kept the cost of servicing government debt in the major advanced economies artificially low, but this can't last forever. As government debt starts to get either downgraded or put on watch by rating agencies, borrowing costs are set to increase.

This combined with pension and health care costs that are set to soar as the population ages will profoundly challenge governments and is likely to lead to considerable political uncertainty and social unrest. Moreover, climbing out of debt will be a major drag on growth and with unemployment rates expected to remain stubbornly high, the adjustment will be painful. The upshot is that standards of living will continue to be squeezed in the advanced economies.

III

Waiting For Sustainable Growth

(Published September 1, 2011)

Three years after the onset of the global financial crisis the G7 economies are still in recovery mode from the global downturn. However, Canada is the only economy in this group that has recouped all the output it lost during the recession. Between 2007 and 2010, Canada's GDP increased at an average annual rate of 0.3 per cent and is now one percentage point above its 2007 peak. It's nothing to brag about for sure, but when viewed against the travails of the other major economies it is certainly an accomplishment.

Despite the return to growth, the level of output in all the other G7 economies remains below pre-crisis levels. Italy's GDP is still 5.3 per cent below its pre-recession level and the economies of the United Kingdom and Japan are both 3.7 percentage points smaller than they were at the start of the recession. And although Germany has had a robust recovery powered primarily by its export sector, its economy is still 0.6 percentage points smaller than it was in 2007 (see table below).

Although Canada has moved from recovery to expansion, its foray into the expansionary phase of the cycle looks like it could be short-lived. Incoming data shows that the economy contracted by 0.4 per cent in the second quarter and two key risks -- the withering of growth in the U.S. and the strong Canadian dollar -- will weigh on the country's economy in the future.

For Canada it will very much depend on whether growth in the United States can pick up speed and here, unfortunately, the odds of that happening do not look good. After averaging growth of 3 per cent in 2010, the pace of recovery in the U.S. has slowed sharply averaging growth of only 0.7 per cent in the first half of this year which is dangerously close to stall-speed.

Aside from the U.S., in the past few weeks the inflow of fresh economic data shows that the recovery in all the G7 countries has started to falter. This deceleration in growth is raising fears that the major economies could again be on the cusp of sliding into another recession.

The problem for governments is that they are running out of policy options to stimulate growth. Indeed, after pursuing a super-loose monetary policy with rock-bottom interest rates and massive doses of quantitative easing, there is precious little that the central banks can do to jump-start growth again. Moreover, as countries press ahead with austerity measures aimed at reigning in unsustainable budget deficits and ballooning government debt levels, this will further darken the outlook for economic growth.

Unfortunately, when dealing with an atypical 'balance sheet' recession, the assumption that expansionary fiscal and monetary policy measures would underpin demand and lead to a self-sustaining recovery has so far proved elusive. Indeed, if anything, it looks like the G7 economies are beginning to seize up again. It's little wonder that the economic forecasts by central banks and investment houses keep getting downgraded at a bewildering rate. It is this uncertainty in the recovery of the real economy that is feeding the wild swings in financial markets. Until some real and sustainable growth takes place we should expect more of the same.

IV

Rise of the Emerging Seven Reshaping the Global Economy

(Published February 1, 2012)

It is now almost three years since the Great Recession ended and profound changes are underway in the world. The global economic axis is shifting fundamentally and a comparison between the major advanced economies of the G7 and the seven largest emerging economies – the E7 – reveals this startling change.

Collectively, the E7 bloc which includes China, India, Indonesia, Brazil, Russia, Turkey and Mexico now accounts for close to 31 per cent of world GDP, up from 19 per cent twenty years ago. During this same time period, the G7 has seen its share of world output fall from 51 per cent to 38 per cent.

It is these divergent trends in growth that have significantly altered the global economic landscape. To put things in perspective, over the four year period from the end of 2007 through to 2011, only four of the G7 economies have regained their pre-recession levels of output. Canada has been the best performer in this group, despite the fact that it is still only 3.1 per cent larger than it was in 2007.

The size of Germany's economy, the second-best performer, is 1.8 per cent larger while the U.S. and French economies have just managed to move ahead of where they were in 2007. The United Kingdom, Japan, and Italy have failed to recover the output lost from the 2008-09 recession. The U.K. economy is 2.6 per cent smaller than it was in the pre-recession peak year of 2007, Japan's is 4.2 per cent smaller, and Italy's is 4.7 per cent smaller

In contrast to the G7 countries, the production of goods and services is larger today in all the E7 economies than it was in 2007. China's economy is 44.6 per cent larger than it was before the crisis and despite a slowing down of growth its GDP is likely to expand by another 8 per cent to 9 per cent this year. Similarly, India's economy is 34.6 per cent larger, Indonesia's is 25.2 cent and Brazil's is 16.5 per cent bigger. Even Mexico's economy, which is 3.9 per cent larger and the E7's worst performer, has outperformed every single member of the G7.

It is now abundantly clear that a sustained recovery remains stubbornly elusive for the G7 economies. Despite massive amounts of monetary and fiscal stimulus, the rate of growth in all of the major advanced economies has been sharply below their respective long-term averages. Moreover, constrained by large debts and deficits, not a single G7 country is expected to achieve growth rates above, or even at, its long-term average for several more years.

In contrast, since 2007, growth in the economies of the E7 has barely deviated from their long-term averages. By 2020 this bloc, given the current trends, will surpass the G7 and account for a greater share of world output. This, in turn, will lead to a shift in the current geo-political power structure. Whether this will be muted or more pronounced remains to be seen.

V

It's Time For A New Approach To Stimulus

(Published May 7, 2012)

When the global financial crisis erupted following the collapse of Lehman Brothers in 2008, the world's major central banks moved quickly to prevent the world economy from falling into a credit-led depression. By sharply lowering interest rates in a series of cuts to record low levels they succeeded in preventing another 1930s style Great Depression from taking hold. But, despite the persistence of ultra-low interest rates since then, growth in the advanced countries has failed to get onto a firmer footing.

Unable to push rates below the zero bound, the world's major central banks have been turning on the electronic spigots and flooding the global financial system with liquidity through a policy of quantitative easing. Since the financial crisis, the U.S. Federal Reserve has more than tripled the size of its balance sheet while the other major central banks have doubled theirs. As a per cent of GDP, the balance sheets of the major central banks now range from 19 per cent in the case of the Federal Reserve (up from 6 per cent in 2007) to 32 per cent for the European Central Bank, up from 17 per cent. Collectively, the balance sheets of these central banks have skyrocketed, rising from $4-trillion in 2007 to over $9-trillion in 2011.

By creating new money to buy up government debt and other financial securities from commercial banks and other financial enterprises, the goal of QE was to boost economic growth, increase inflation, and lower the unemployment rate. However, when one examines the impact of quantitative easing, it is apparent that the policy has misfired. After an initial spurt, economic growth has started to slow sharply in the advanced countries, job creation has been weak, and the unemployment rate is still well above pre-recession levels.

The problem is that QE is a very indirect way of trying to stimulate the economy. While QE has helped to recapitalize banks, which continue to focus on repairing and purging their balance sheets from toxic assets, it has done little to help the flow of credit to the real economy. Despite a super-accommodative monetary policy, credit conditions still remain tight particularly for small and mid-sized companies.

Trying to force feed economic growth by saturating the system with trillions of dollars when households and governments are deleveraging has proved to be an ineffective way to spur the recovery. A better approach would be for the central banks to intervene directly into the economy by lending money to small and mid-sized companies who in the main are the lead generators of employment.

Although the whole gamut of monetary policy needs a re-think, central bankers appear determined to stay the course. Indeed, the Federal Reserve has indicated that it intends to keep rates at near zero levels until well into 2014. And, there is talk that they may be preparing for another round of QE in the (vain) hope that flooding the system with even more money will eventually lead to an upsurge in growth.

Obviously, no one is paying heed to Japan's decades long slump. More's the pity.

VI

IMF Projections Question G7's Commitment to Tackling Debt

(Published September 24, 2012)

When the Great Recession wreaked havoc on the public finances of major advanced economies, the standard narrative was that governments would move to rein-in spending. The latest IMF projections, however, might prompt a rewrite. The debt-to-GDP ratios in all the G7 economies are projected to be considerably higher in 2017 – a full ten years after the onset of the crisis.

An analysis of the IMF's latest medium-term projections shows that over the 2007-2017 period the level of debt is projected to increase at a faster pace than growth in output in each of the major economies further pushing up the debt-to-GDP ratio. In 2007, debt ratios ranged from a low of 44 per cent in the United Kingdom to a high of 183 per cent in Japan. But, with the level of debt projected to grow faster than their respective economies, the debt-to-GDP ratio is slated to be higher in all G7 nations in 2017 compared to 2007. By 2017, the ratio is expected to range from a low of 71 per cent in Germany to a high of 257 per cent in Japan

In nominal terms the total value of goods and services produced is projected to be larger in all the G7 countries in 2017 compared to a decade earlier, but growth rates are expected to vary widely. For instance, the IMF projects that by 2017 Japan's economy will only be 1 per cent larger than it was in 2007 whereas at the other end of the spectrum, Canada's economy will be 46 per cent bigger, the U.K.'s 42 per cent and the United States at 40 per cent. The major European economies are also projected to grow but at a considerably slower pace. For example, France's economy is estimated to be 31 per cent larger followed by Germany at 24 per cent and Italy at 12 per cent.

Since 2007, national debt levels have been exploding in all the major advanced economies. By 2017, the level of government debt is expected to more than double in the US, rising from $9.4-trillion dollars in 2007 to $22.3-trillion. The debt outlook for the UK is even more dramatic. That country's national debt is projected to hit £1.7-trillion in 2017, a jump of £1.1-trillion from its 2007 level. Canada's debt level is expected to top $1.6-trillion in 2017, up from about $1-trillion in 2007. In France, Germany and Italy, debt levels are projected to top €2-trillion in each country and Japan is expected to tack on another ¥39-trillion in debt.

With public finances that are already at abysmal levels, the projected deterioration in the debt levels of all the major advanced economies will continue to challenge the ability of governments to service these debt levels. The best way out of the debt trap is via economic growth but despite letting loose vast amounts of monetary and fiscal stimulus growth has failed to gain traction. Indeed, the latest data shows that GDP is heading down not only in the G7 economies but it is also decelerating sharply globally.

The bottom line is that winding down these mountainous debt levels will take a long time and will act as a brake on economic growth. It looks like this decade will be the decade of the 'Great Stagnation' in the G7 economies.

About The Author

Ranga Chand is an international economist and financial author. He was born in Uganda in 1946 but grew up in England. He was educated at King's School Worcester where he was a boarder from 1958 to 1964. On leaving King's Ranga studied economics and obtained an MA from the University of Toronto in 1970. After his graduation he spent his entire working career in Canada.

He has held senior positions with Canada's Department of Finance and served as a director of the Conference Board of Canada before joining a major stock brokerage firm. He has represented Canada at numerous economic forums, including the OECD in Paris, the United Nations, and the Kiel Institute for the World Economy in Germany.

At the invitation of Robert (Bob) Mundell, the Nobel Laureate, Ranga also taught economics at the University of Waterloo, Ontario, Canada from 1973 to 1974.

From 2000 to 2003 he hosted the popular television show "Talking Mutual Funds with Ranga Chand" which aired weekly on Canada's Report on Business Television (ROBTV) and reached over 4 million viewers nationwide. He has published extensively in the field of economics, is the author of a number of best-selling investment books and was also a regular contributor on economic and financial issues to Canada's premier national newspaper The Globe and Mail.

Ranga and his wife, Sylvia, now live in Canterbury, England.

BOOKS BY RANGA CHAND

(1994)    (1994)

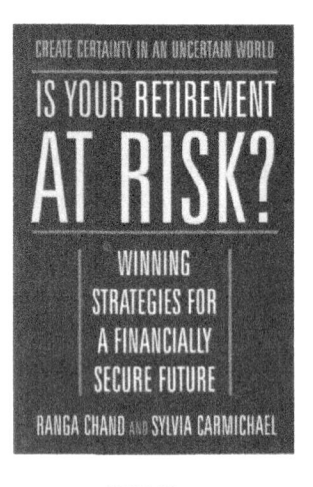

(2002)    (1999)

BOOKS BY RANGA CHAND

(1998)

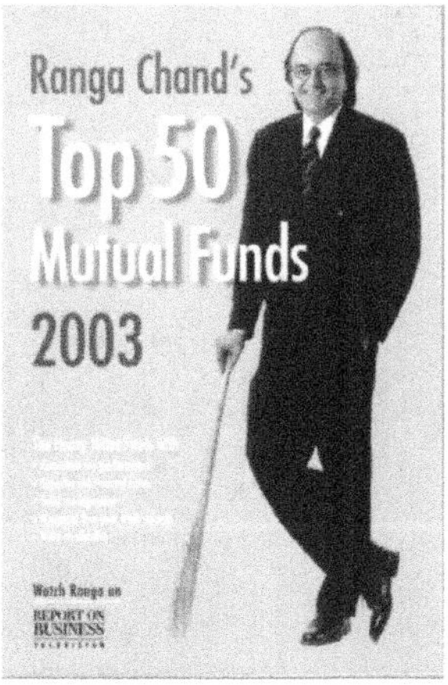

(2002)

(1996)

(1997)

Made in United States
Orlando, FL
20 May 2024